NUTRITION AND DIET RESEARCH PROGRESS

GENISTEIN AND DAIDZEIN

FOOD SOURCES, BIOLOGICAL ACTIVITY AND HEALTH BENEFITS

NUTRITION AND DIET RESEARCH PROGRESS

Additional books in this series can be found on Nova's website
under the Series tab.

Additional e-books in this series can be found on Nova's website
under the e-book tab.

NUTRITION AND DIET RESEARCH PROGRESS

GENISTEIN AND DAIDZEIN

FOOD SOURCES, BIOLOGICAL ACTIVITY AND HEALTH BENEFITS

MASAYOSHI YAMAGUCHI
EDITOR

nova publishers
New York

Library of Congress Cataloging-in-Publication Data

ISBN: 978-1-63483-194-9
Library of Congress Control Number: 2015942937

Published by Nova Science Publishers, Inc. † New York

CONTENTS

PREFACE

Soybeans are greatly used in many foods, such as tofu, soy milk and yogurt, and it plays an important role as a functional food in the health and prevention of various diseases. Soybeans contain rich isoflavones including genistein, daidzein and glycitein, and both genistein and daidzein occur at higher concentrations. Isoflavones are structurally and functionally similar to 17ß–estradiol. Isoflavones are found to bind to estrogen receptors and the peroxisome proliferator-activated receptor (PPAR) family and aryl hydrocarbon receptors. Moreover, isoflavones affect activity of tyrosine kinases, histones, transcription factors, adenosine mono phosphate pathways, RNA expression and other intracellular regulators. Thus, isoflavones play a multifunctional role in cell regulation. This book focuses on recent advances in the food sources, biological activity and health benefits of genistein and daizen.

Chapter 1 – Soybeans as a whole or their food isolates are found in many foods, such as tofu, soy milk and yogurt. In women, the consumption of these foods is especially associated with beneficial health effects, such as menopause vasomotor symptoms reduction, osteoporosis prevention, genital atrophy mitigation, and breast cancer attenuation. Among them, some elements have nonsteroidal structure and can bind to estrogen receptors, which are classified as phytoestrogens. Isoflavones are phytoestrogens with high biological activity, especially those named genistein and daidzein. Both bind to estrogen receptors with greater affinity to β receptors. Genistein (4', 5,7 trihydroxyisoflavone) is considered a potent anticancer agent due to its capacity of inhibiting tumor cell growth factors in several tissues and diseases, such as breast, prostate, and neck cancers as well as melanoma and leukemia. Moreover, they block the cell cycle and induce cell death by apoptosis, inactivating NF-kB and activating caspase-3 in prostate cancer cells. In addition, they act as potent angiogenic factors and tumor cell proliferation inhibitors. Some studies suggest that genistein inhibited the growth of tumor cells due to inhibition of gene expression of protein tyrosine kinase of topoisomerases I and II in different cell types. Other studies also showed that genistein was capable to prevent apoptosis of normal cells through their antioxidant properties. Daidzein (7,4-dihydroxy isoflavone), another isoflavone, when ingested is metabolized in equol by intestinal bacterias. It has been hypothesized that production of equol is the key to the effectiveness of isoflavone, which brings health benefits. Such benefits were reported on renal function in hypertensive postmenopausal women, as well as in osteoporosis and prostate cancer prevention. Taking together, clinical, experimental and cell cultures studies suggest that isoflavones, mainly genistein and daidzein, play an important role in preventing chronic diseases, such as osteoporosis, cancer and diabetes. Thus, more information about isoflavones as a non-medicinal alternative has become extremely important because of the high incidence of chronic diseases worldwide.

Chapter 2 – Oestrogen replacement therapy is not thought to be a safe treatment for prevention of cardiovascular disease in menopausal women; isoflavones are a possible alternative. The early stage of atherosclerosis is characterized by increased levels of inflammatory cytokines, oxidative stress, and the expression of adhesion molecules (CAMs) on vascular endothelium. The endothelium produces nitric oxide (NO), a vasoprotective and anti-atherosclerotic agent. Oestrogen has beneficial effects on the cardiovascular system by enhancing NO production. Oestrogen-like compounds such as isoflavones are also suggested increase NO production, exerting anti-atherogenic effects. Therefore, compounds that suppress CAMs, reduce oxidative stress or increase NO may be usable as therapeutic agents against the development of atherosclerosis in menopausal women. Soy isoflavones exert anti-atherogenic effects. Isoflavones are present in soy foods mainly as glucosides. Soy isoflavone aglycones, the biological active estrogenlike compounds, are absorbed faster and in higher amounts than their glucoside derivatives. In addition, the aglycone forms have a higher biological activity implying that isoflavone aglycone-rich products may be more effective than the ones rich in glycosides in preventing chronic diseases such as coronary heart disease. Objective: In this chapter the authors review and compiled all the results already published and present some unpublished results about our study comparing the effects of an extract of soybean fermented by *Aspergillus awamori* (a β-glucosidases-producing filamentous fungi) on which polyphenol glucosides were bioconverted to aglycone forms (BSE) with a non-fermented extract (SE) on adhesion molecules expression (VCAM-1, ICAM-1 and E-selectin), oxidative stress induced by hydrogen peroxide and on production of nitric oxide, prostaglandin E2 and Endothelin-1 i*n vitro* in human endothelial cells (HUVEC). Results: Bioconverted soybean extract (BSE) decreased VCAM-1, ICAM-1 and E-selectin on cell surface and on cell culture supernatant, decreased the oxidative stress induced by hydrogen peroxide and enhanced endothelin-1, nitric oxide and prostaglandin E2 production, while the unfermented extract also decreased adhesion molecules expression and oxidative stress; enhanced endothelin-1 production but was not able to increase nitric oxide and prostaglandin E2 production. Conclusion: The aglycone -rich soybean extract was able to reduce adhesion molecules expression and oxidative stress, being more effective to reduce sICAM-1 expression and only the aglycone-rich forms soybean extracts was able to increase nitric oxide and prostaglandin E2 production, demonstrating that in edothelial cells *in vitro*, may be usable as therapeutic agents against the development of atherosclerosis.

Chapter 3 – Bone homeostasis is regulated through osteoclasts and osteoblasts. Osteoporosis, which is induced with its accompanying decrease in bone mass with increasing age, is widely recognized as a major public heath problem. Bone loss may be due to decreased osteoblastic bone formation and increased osteoclastic bone resorption. There is growing evidence that nutritional and food factors may play a part in the prevention of bone loss with aging and have been to be worthy of notice in the prevention of osteoporsis. Genistein, which is contained in soybeans, has been shown to have a stimulative effect on osteoblastic bone formation and a suppressive effect on osteoclastic bone resorption, thereby increasing bone mass. This factor has a regulative effect on protein synthesis and gene expression, which are related to bone formation in osteoblastic cells and bone resorption in osteoclastic cells. Oral administration of genistein has been shown to prevent on bone loss in ovariectomized rats, an animal model for osteoporosis. Moreover, intake of dietary soybean genistein has been shown to have a preventive effect on osteoporosis in human subjects. These findings have been demonstrated a role of genistein in the prevention of bone loss with aging. This chapter introduces a role of dietary soybean genistein in osteoporosis prevention.

Chapter 4 – Phytoestrogens are natural polycyclic phenols found in certain plants that are structurally and functionally similar to 17ß–estradiol. When ingested and metabolized, they bind to estrogen receptors (mainly estradiol $ß_2$ receptors – ER $ß_2$) and produce estrogenic effects. Furthermore, they also bind to peroxisome proliferator-activated receptor (PPAR) family and aryl hydrocarbon receptors. Moreover, phytoestrogens affect activity of tyrosine kinases, histones, transcription factors, AMP (Adenosine mono phosphate) pathways, RNA expression and other intracellular regulators. Current research demonstrates that phytoestrogens are helpful in reducing the intensity and frequency of hot flushes in menopausal women. Still, complete elimination cannot be achieved, as phytoestrogens have weaker biological activity than endogenous and synthetic estrogens. Certain phytoestrogens have also been shown to decrease vaginal atrophy. Moreover, unlike hormone therapy, phytoestrogens were not found to increase clotting risk. Studies confirmed beneficial effects of phytoestrogens on endothelial cells, vascular smooth muscle and extracellular matrix cells. Due to nitrogen oxide (NO) production they decrease arterial stiffness and exert anti-atherosclerotic effects. Phytoestrogens also affect angiogenesis and reduce inflammation and tissue damage, which delays the progression of atherosclerosis. Furthermore, phytoestrogens increase high density lipoprotein (HDL) and decrease low density lipoprotein (LDL) concentrations in human plasma. Neuroprotective effects of phytoestrogens are related to both their antioxidant properties and interaction with ER. Literature data show that phytoestrogens might improve cognition, prevent or at least postpone dementia. By activating PPARγ receptors phytoestrogens down-regulate pro-inflammatory cytokines, such as Cyclooxygenase 2 (COX-2) and Inducible nitric oxide synthase (iNOS), increase reverse cholesterol transport and insulin sensitivity. Phytoestrogens increase energy expenditure by affecting AMP activated kinase signaling cascades, which inhibit adipogenesis. Therefore, they can exert different effects on metabolism and may complement conventional treatment for diseases linked to metabolic syndrome, polycystic ovary syndrome and diseases related to hyper-androgenism and obesity. Phytoestrogens have shown antiresorptive effects on bone through ER signaling. Additionally, they also affect broad spectrum of molecular mechanisms supporting osteogenesis by inhibiting PPARγ, Sirtuin 1 and CCAAT-enhancer-binding proteins (C/EBP) α, the key adipogenic transcription factors. Stimulating main osteogenic transcription factors Core-binding factor subunit alpha-1 (CBF-alpha-1) and Osterix, they present themselves as a new strategy for treating bone loss. Literature data strongly support the relationship between phytoestrogen diet supplementation and the lower risk for some cancers. Phytoestrogens have been shown to inhibit the carcinogenesis through the modulation of genes that control the cell-cycle progression. They inhibit the activation of the nuclear transcription factor, kappa light polypeptide gene enhancer in B-cells (NF-kappa B) and protein kinase B signaling pathway, which are known to maintain a homeostatic balance between cell survival and programmed cell death (apoptosis). Phytoestrogens are also implicated in reversion of epigenetic events observed in some cancers. Furthermore, phytoestrogens are known to have potent antioxidant properties (inhibition of oxidative stress) and are found to be potent inhibitors of angiogenesis and metastasis. Moreover, phytoestrogens regulate mucosal immune response by suppressing dendritic cell function. In conclusion, phytoestrogens appear to exert numerous positive health effects and consequently are recommended for use in various clinical fields.

Chapter 5 – Bone remodeling consists of a balance between bone formation by osteoblasts and bone resorption by osteoclasts. Osteoporosis is the result of increased bone resorption and decreased bone formation causing a decreased bone mass density, loss of bone microarchitecture, and an increased risk of fractures. Osteoporosis is most common in postmenopausal women, primarily due to the effects of hypoestrogenemia. In this chapter, the efficacy and mechanisms of action of phytoestrogens in promoting osteogenesis and the use of phytoestrogens as potential therapeutic agents for osteoporosis will be presented. Two phytoestrogens that have gained significant attention are daidzein and genistein. Overall, daidzein and genistein share many cellular effects to reverse the phenotypes observed in ovariectomized animals. Studies have shown that genistein effectively reduces the impacts of osteoporosis, while efficacy studies investigating daidzein have shown mixed results, which will be discussed. Several studies showed that daidzein is more efficient than genistein in preventing ovariectomy-induced bone loss in rats. With regard to the molecular signaling that induces these phenotypic cell-specific effects, daidzein and genistein both act primarily through estrogen receptor dependent mechanisms at lower concentrations but may also act through different pathways at higher concentrations, which will be presented in this chapter. The importance of metabolites of daidzein and genistein and other phytoestrogens that may play a role in osteogenesis will also be addressed.

Masayoshi Yamaguchi, PhD, IOM, FAOE, DDG, DG,
Adjunct Professor
Department of Hematology and Biomedical Oncology
1365 C Clifton NE,
Atlanta, Ga, 30322
Yamamasa1155@yahoo.co.jp

In: Genistein and Daidzein
Editor: Masayoshi Yamaguchi
ISBN: 978-1-63483-194-9

Chapter 1

EFFECTS OF GENISTEIN AND DAIDZEIN IN FEMALE REPRODUCTIVE TRACT

Adriana Aparecida Ferraz Carbonel[1,*], Luciana de Oliveira Bruno[2], Cristiane de Paula Teixeira[1], Ricardo Santos Simões[3], João Henrique Castello Girão[2], Manuel de Jesus Simões[1] and Clélia Rejane Antonio Bertoncini[4]

[1]Department of Morphology and Genetic, Histology and Structural Biology Division, São Paulo Federal University (UNIFESP/EPM), Brazil
[2]Department of Gynecology, Gynecology Division, São Paulo Federal University (UNIFESP/EPM), Brazil
[3]Department Obstetrics and Gynecology, Gynecology Division, Faculty of Medicine, University of São Paulo (FMUSP), Brazil
[4]Center for the Development of Experimental Models for Medicine and Biology (CEDEME), São Paulo Federal University (UNIFESP/EPM), Brazil

ABSTRACT

Soybeans as a whole or their food isolates are found in many foods, such as tofu, soy milk and yogurt. In women, the consumption of these foods is especially associated with beneficial health effects, such as menopause vasomotor symptoms reduction, osteoporosis prevention, genital atrophy mitigation, and breast cancer attenuation. Among them, some elements have nonsteroidal structure and can bind to estrogen receptors, which are classified as phytoestrogens. Isoflavones are phytoestrogens with high biological activity, especially those named genistein and daidzein. Both bind to estrogen receptors with greater affinity to β receptors.

Genistein (4', 5,7 trihydroxyisoflavone) is considered a potent anticancer agent due to its capacity of inhibiting tumor cell growth factors in several tissues and diseases, such as breast, prostate, and neck cancers as well as melanoma and leukemia. Moreover, they

* Corresponding author: Adriana A. F. Carbonel. E-mail: adricarbonellfisio@hotmail.com.

block the cell cycle and induce cell death by apoptosis, inactivating NF-kB and activating caspase-3 in prostate cancer cells. In addition, they act as potent angiogenic factors and tumor cell proliferation inhibitors. Some studies suggest that genistein inhibited the growth of tumor cells due to inhibition of gene expression of protein tyrosine kinase of topoisomerases I and II in different cell types. Other studies also showed that genistein was capable to prevent apoptosis of normal cells through their antioxidant properties. Daidzein (7,4-dihydroxy isoflavone), another isoflavone, when ingested is metabolized in equol by intestinal bacterias. It has been hypothesized that production of equol is the key to the effectiveness of isoflavone, which brings health benefits. Such benefits were reported on renal function in hypertensive postmenopausal women, as well as in osteoporosis and prostate cancer prevention. Taking together, clinical, experimental and cell cultures studies suggest that isoflavones, mainly genistein and daidzein, play an important role in preventing chronic diseases, such as osteoporosis, cancer and diabetes. Thus, more information about isoflavones as a non-medicinal alternative has become extremely important because of the high incidence of chronic diseases worldwide.

INTRODUCTION

Menopause is characterized by reduced levels of estrogens which trigger uncomfortable symptoms such as heat waves, night sweats, sleep disorders, psychological disturbances and vaginal dryness among others. Menopause symptoms significantly affecting the quality of life, and it has been reported that heat waves are the most annoying trouble (Avis et al., 2001). To alleviate the symptoms it is recommended the hormone therapy (HT). However, a multicenter study (WHI) showed that the benefits of hormone therapy with estrogen alone could not be so inestimable, since there is increased risk of stroke and venous thromboembolism (VTE). Moreover, the combination of estrogen with progesterone may also increase the risk of causing breast cancer and heart attack (Rossouw et al., 2002). As a result, the role of HT was limited, recommending lowest doses of hormones for short period to treat postmenopausal symptoms. Therefore new drugs are being evaluated to try to reduce menopausal symptoms (Chen et al., 2014).

Population studies show that Asian women present less menopause symptoms and breast cancer rate than Caucasians (Ricciotti et al., 2005; Khaodhiar et al., 2008; Patisaul et al., 2010; Aso et al., 2012; Jenks et al., 2012). When the studies were related to feed consumption per quantify of soy per quantity of soy, the result was a dose-response relationship with a statistically significant trend for decreased risk of breast tumor, which means a reduction in risk of 16% for each 10 mg of isoflavones per day consumed (Wu et al., 2008). Studies evaluating the adoption of western diets for soybean asian immigrants or lower intake rates have found increased risk of cancer in breast and prostate, suggesting that dietary and environmental factors are more important than the racial characteristics (Morton et al., 2002; Morimoto et al., 2014).

Soybean was always valued in the East as a nutritious food because it contains high concentrations of protein substances, lipids, vitamins and minerals. Soybean cultivation in the central China dates back over 5,000 years. The "tofu", one made of soy food, has been known since the second century BC, and was introduced by Korean monks in Japan in the seventh century BC. Furthermore, the use of soya beans and its derivatives as a food source is part of the eastern culture, as well as its use in the prevention and treatment of diseases has a long tradition in Chinese medicine (Barnes et al., 1994).

Thus many western women began to consume soy-based foods, or taking medicines containing the active ingredients of soy, to relieve postmenopausal symptoms or prevention of breast cancer. However the impact of soy intake or its active principles on the prevention of breast cancer is not yet clearly understood.

Some articles show that isoflavones have pleiotropic effects, including preventive and even curative action of chronic diseases, colon, breast and prostate cancers, cardiovascular and hormone-dependent diseases, obesity, metabolic syndrome and menopausal disorders such as osteoporosis (Somekawa et al., 2001; Liu et al., 2004; Messina, 2014). However there are articles that point out that soy has no effect on menopausal symptoms. So, we tried to assess in this chapter some aspects of the soybean action on the female reproductive system.

It is widely known that soy contains substances whose molecules have hormonal and pharmacological actions.

They are called isoflavones and the main are named genistein, daidzein and glycitein the (Adlercreutz et al., 2004; Mateos-Aparicio et al., 2008). Isoflavones are chemical compounds structurally and/or functionally similar to estrogens (phytoestrogens) or their active metabolites. They are present in a large variety of plants, especially in cereals. In leguminous crops, case of soybeans, the isoflavonas are present in relatively high concentrations (Zhao and Mu, 2011).

Several studies have shown that phytoestrogens act on some biological mechanisms such as proliferation, differentiation, and protein synthesis in various target cells. These effects are dependent on the type and content of the receiver under which they operate and the target cell. They can exert beneficial effects on the human organism, such as therapeutic and preventive actions in carcinogenesis, atherosclerosis, osteoporosis, and improvements in the symptoms of menopause. However, many experimental and clinical studies examining the impact of the consumption of soybean phytoestrogen compounds show contradictory results.

Due to the divergences among the various studies that evaluated the effect of isoflavones, we should mention that these phytohormones are used in small concentrations and can be altered by many factors such as type of seed, cultivation, storage and extraction methodology. It is known that the extractions performed with alcoholic solvents remove most of the isoflavones from the material, which may result in large differences in the concentration of the active ingredients present in the commercial products. Thus it is possible to obtain extracts with isoflavones concentrations ranging from 0.5% to 90% of the total weight, and also differences in proteins and other components content, which may explain the many divergences when the therapy assessments are confronted (Messina, 2014; Bedell et al., 2014). An additional concern is the possibility of the phytoestrogens endangers the health of certain age groups, especially menopausal women seeking soybeans or derived hormonal alternative treatment (Patisaul and Jefferson, 2010; Chen et al., 2011; Bedell et al., 2014).

Isoflavones Contained in Soybean

The naturally occurring isoflavones in soy seeds are genistein (4,5,7-trihydroxy isoflavone), daidzein (4,7 - dihydroxyisoflavone) and glycitein (4,7-dihydroxy-6-

methoxyisoflavone), and both genistein and daidzein occur at higher concentration (Dixon and Ferreira, 2002; Di Virgilio et al., 2004; Wegrzyn et al., 2010).

These molecules have a non-steroid structure, which contains a phenolic ring. This ring enables the connection of isoflavones to estrogen receptors (Pilšáková et al., 2010). The two benzene rings connected by a heterocyclic pyran ring characterizing the structure of these compounds (Ko et al., 2014).

The food consumed soy products typically contains 0.01-0.3% glucose conjugated isoflavones, genistein which means (GE), daidzein (DE) and glycitein (GLYE) as gliconas. In this form they are not absorbed by the human body since only free isoflavones without the sugar molecule, called the aglycones, are able to cross the cell membrane. Thus after having been consumed, the isoflavone glucosides are readily hydrolyzed by β-glucosidases present at the brush border of the enterocytes or produced by bacteria present in the intestine, in bioactive corresponding aglycones (genistein, daidzein and glycitein).

After ingestion of soy-based foods, we have found a biphasic pattern in the appearance of isoflavones in the plasma as well as urine, which suggests that it is due to their metabolism in the small intestine and absorption during the first 2 hours (about 10%); and after 4-6 h in the large intestine (about 90%). As daidzein (DE) and genistein begin to be absorbed minutes after ingestion, equol (EQ) appears in plasma only 8h after ingestion due to the transit time up to the colon, where conversion occurs from DE to EQ by the intestinal microbiota (Franke et al., 2014).

After absorption, these molecules are incorporated into chylomicrons that carry them into the lymphatic system and then to the circulatory system. Thus isoflavones are distributed in all extrahepatic tissues, where they will exercise their metabolic effects before returning to the liver to be metabolized and eliminated. Studies in human plasma shows that only EQ occurs predominantly as an aglycone (1%) while the others appear predominantly as glucuronides (75%) and sulfate (24%) (Gu et al., 2006; Franke et al., 2014) (Figure 1). Equol is a metabolite of daidzein, not found in plant foods, made exclusively by the action of intestinal bacteria. In order to occur the bioconversion of daidzein in its metabolite equol, a very specific intestinal microorganism is needed, and this bioconversion only occurs in 30 to 50% of the people. However, it is known that equol is produced under anaerobic conditions, and further that vegetarians and Asian origin individuals are at greatest ability to produce equol. Some authors also report that the genesis of equol may be critical for the benefits associated with a diet rich in soy, because this molecule has a higher estrogenic activity (10 to 100 times higher) compared to genistein (Setchellet al., 2005; Patisaul and Jefferson, 2010).

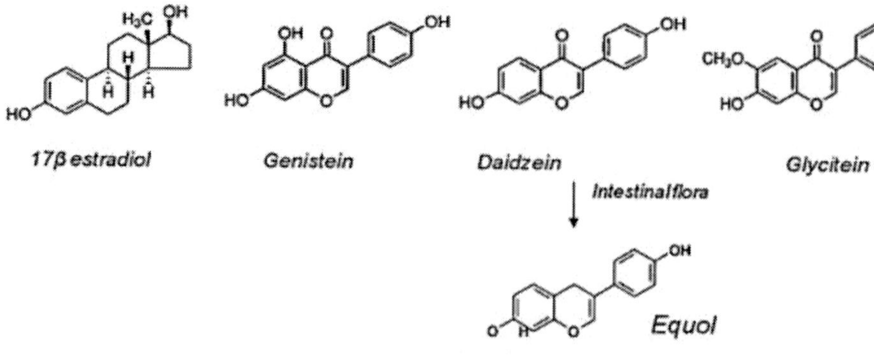

Figure 1. Molecular Structure of 17β estradiol, genistein, daidzein and equol.

They are not clearly elucidated what are the factors that influence the biotransformation capacity of daidzein to equol, but genetic factors, intestinal physiology and diet should contribute to interindividual differences (Patisaul and Jefferson, 2010). The biological efficiency of this substance is due to its greater affinity for estrogen receptors, anti-androgenic property and potent antioxidant activity.

Population studies evidence a direct association between urinary excretion of equol and reduced risk of cancer in breast, endometrium and prostate, as well as attenuation of menopause symptoms, hypoglycemic activity, and prevention of cardiovascular diseases (Xu et al., 2015). When daidzin, glycitin are transported to the liver, they are conjugated by liver enzymes and suffer sulfation, glycosylation and glucuronidation, so that they become very soluble in water (Gu et al., 2006).

Genistein is transformed into dihidrogenisteína-hydroxy and 6'-O-DMA, which is degraded to 4-hydroxyphenyl-2-propionic acid (Di Virgilio et al., 2004; Usui, 2006; Patisaul and Jefferson, 2010). The excretion of conjugated metabolites occurs mainly in the urine, whereas the unconjugated forms are excreted in the feces (Klein and King, 2007).

Thus the intestinal microflora, the administered dose and the intestinal transit time plays an important role in metabolism, bioavailability, biological activity and metabolomic profiles of isoflavones in the diet, which may help to explain the differences regarding their efficacy (Izumi et al., 2000; Zubik and Meydani, 2003; Setchell et al., 2003; Cederroth and Nef, 2009; Zheng et al., 2013).

Molecular Mechanisms of Action of Soy Isoflavones

As mentioned soy isoflavones, genistein, daidzein and glycitein, present a structural similarity to 17β-estradiol, which explains its estrogenic activity (Figure 1). The structural and functional similarity of soy isoflavones with estradiol allows these molecules to bind to the estrogen receptor (ER) alpha and beta, although they present lower affinity (Hwang et al., 2013). The binding affinity of genistein is 4% to ERα and 87% to ERβ (Morito 2001; Cavallini et al., 2008; Lima et al., 2014). When compared with the affinity of 17β estradiol, genistein's affinity to ERβ is about 20-30 times greater than that to ERα. The distribution of ERα and ERβ varies in different tissues, for example; in the female reproductive tract, the uterus and breast, ERα are abundant, whereas in the bone and prostate ERβ predominates. This difference in the distribution of α and β receptors leads to different effects of isoflavones among the various tissues (Smith et al., 2004; Jordan, 2007; Oseni et al., 2008). Apart from exerting effects on ER, recent studies also indicates that phytoestrogens exert an action through membrane G-protein coupled receptor (GPR30). This binding induces a more rapid cellular response involving the activation of fosfatidilnositol 3-kinase (PI3K), calcium mobilization and increased production of adenosine 3', 5'-cyclic monophosphate (cAMP) (Prossnitz et al., 2011; Luo et al., 2012; Kajta et al., 2013). In addition to their estrogenic activity, there is evidence that isoflavones can modulate the expression of some genes that control cell survival (Moiseeva et al., 2007), cell cycle (Touny et al., 2006) and apoptosis (Chen et al., 2015). They exhibit antioxidant properties (Teixeira et al., 2014; Xu et al., 2015) and inhibit the tyrosine kinase activity to disrupt cellular pathways (Akiyama et al., 1987; Baechler et al., 2014; Li et al., 2014), suppress nuclear factor KB (NFKB) and regulate protein kinase B (AKT), as well as other MAPK signaling pathways (Li et al., 2012; Yu et al., 2012). Currently it is possible to obtain pure extracts of isoflavones, so we begin to deal with each of them.

GENISTEIN

Genistein has a structural similarity with 17β-estradiol, having a molecular weight as well as a similar hydroxylation pattern.

It may bind to the receptors (ERα) and (ERβ) or membrane receptor (GPR30), regulating the expression of genes (Jian, 2009). The binding affinity of this isoflavone for ERβ is higher than for ERα and there is a competition between genistein and estradiol-17β (Banerjee et al., 2008; Pavese et al., 2010). Genistein inhibits the proliferation of cancer cells through a blockade of cell cycle progression at G2 / M phase (Hewitt and Singletary, 2003). This isoflavone also inhibit nuclear factor-kB (NFkB), a transcriptional factor that regulates pathways of the inflammatory response and cell proliferation, which can induce apoptosis in breast cancer cells and prostate (Jian, 2009; Hsu et al., 2010).

It is also known as a potent inhibitor of protein tyrosine kinase (PTK) and topoisomerase II in various cancer cells; it decreases oxidative damage of DNA and tumorigenesis. A PTK catalyses the phosphorylation of tyrosine residues of tumor cells and other proteins, including growth factors involved in tumor cell proliferation (Fitzpatrick, 2003; Hewitt and Singletary, 2003). Genistein also regulates the PTK in the brain, with a neuroprotective effect (Patisaul and Jefferson, 2010). Preclinical studies showed that genistein may also inhibit tumor angiogenesis and metastasis, exerting effects on the early stage primary tumor growth as well as in the final steps of the metastatic cascade. These effects were reported in various targets of metastatic cascade, which are essential for cell adhesion and invasion process.

Clinical trials confirm the results obtained from preclinical studies, since a decrease in advanced metastatic disease markers has been established, such as MMP-2 (matrix-metalloproteinase 2) and PSA (prostate specific antigen). These effects were observed after consumption of low concentrations from genistein, which can be reached through the daily intake of soybean seeds (Pavese et al., 2010).

Researches on the effect of genistein into combination with tamoxifen tumor cell line human breast adenocarcinoma (MCF7) in rats showed that they act synergistically to inhibit tumor proliferation and inducing apoptosis of cancer cells. Modulation of the levels of estrogens and their biological activity is a mechanism in which the combination of genistein/ tamoxifen exerts its effect on tumor growth (Mai et al., 2007).

Studies in rats have shown that genistein administered either immediately or later result in lower risk of cancer when compared to estrogen alone (Carbonel et al., 2014).

DAIDZEIN

The daidzein is considered a non-toxic compound capable of inducing cell death in several cancer cell types (Jin et al., 2010). Studies have shown that daidzein interrupts the cell cycle of human breast cancer cells in both G1 and G2/M as well as significantly increased activity of caspase-9.

As the activity of caspase-9 increases, daidzein induces apoptosis through the intrinsic pathway regulated by cellular signaling in mitochondria (Choi and Kim, 2008). However, despite the claim that the anticancer activity of daidzein is mediated through cell cycle regulation and apoptosis, the specific mechanisms are not well understood (Jin et al., 2010).

The daidzein appears to inhibit cell viability in a time and dose-dependent manner (Choi and Kim, 2008). This isoflavone inhibits growth of cancer cells ER (+) and ER (-) in different ways. Studies indicate that the antiproliferative effect on ER (+) cells shall be subject to a phenomenon of saturation of ER receptors, which does not occur in the ER cells (-). Thus high concentrations of daidzein may be beneficial in cancer chemoprevention of ER (-) (Guo et al., 2004).

EQUOL

As mentioned above, the estrogenic action of isoflavones is not only due to their ingested conjugated, but also to biologically active metabolites. The daidzein when converted into equol exerts estrogenic activity that exceeds that of the genistein (Morito et al., 2001; Setchell et al., 2002). Some studies report that isoflavone intake may stimulate the growth of tumor cells in the breast (Hshie et al., 1998; Ju et al., 2006). However, data from clinical trials using pure genistein did not have clearly demonstrated that it produces deleterious effects (Wu et al., 2013). A mechanism by which isoflavones can stimulate the development of breast cancer is its affinity associated with competition with endogenous estrogens.

Nevertheless, it has been shown that isoflavones have anti-proliferative, proapopticas, anti-angiogenic, anti-oxidants and anti-inflammatory properties (Akiyama et al., 1987; Watanabe et al., 2000).

EFFECTS OF SOY ISOFLAVONES ON THE FEMALE REPRODUCTIVE SYSTEM

There are doubts about the beneficial effects of isoflavones on the female reproductive system and its annexes, in particular in the breasts. Thus the scientific evidence regarding the beneficial effects of the consumption of isoflavones in the prevention and treatment of side effects of menopause are based on experimental and epidemiological studies.

Prospective Cohort Study showed the higher the intake of isoflavones the greater is the protection against breast cancer in hispanic women, african, japanese and american women (Morimoto et al., 2014). However, a study examining the effects of soy isoflavones on genomic pathways related to breast cancer showed that a high plasma concentration of genistein induces overexpression of genes that drive the cell proliferation cycle. Such result raised the concern that a subgroup of women who consume soy at high concentrations can be adversely affected by the expression of genes that drive breast cancer (Shike et al., 2014).

In vitro studies have reported that high concentrations of isoflavones reduce cell proliferation, whereas low concentrations exert stimulatory effects. Low concentrations of isoflavones also have the capacity to modify the expression of genes critical for cell survival (Moiseeva et al., 2007), cell cycle control and apoptosis (Touny et al., 2006).

Laboratory studies have shown that genistein and daidzein have the ability to inhibit tyrosine kinase (PTK) thereby preventing the signal path between the growth factor, its receptor, and the DNA, so that does not occurs neither the activation of cell proliferation nor the angiogenesis (Tepper et al., 2007; Li et al., 2008).

It is well known that the cell cycle is regulated at different stages by cyclins, cyclin-dependent kinases (CDKs) and cyclin-dependent kinase inhibitors (CDKIs). Thus, these regulators are important targets for cancer prevention and therapy. In this context, a study evaluating the combination of genistein and daidzein revealed increased p53 and decreased cyclin B1rr protein expression (Choi et al., 2000).

From experimental study in rats using different doses of isoflavones we found that they do not stimulate proliferation of mammary proliferation, and they have a protective effect due to decreased ability to bind to the estrogen receptor alpha (ERα) (Carbonel et al., 2011). When we studied the effects of soy isoflavones on the expression of genes that control cell growth, we note blockage of the cell cycle and possibly for cancer prevention (Carbonel et al., 2014). Despite the numerous studies either for or against the indication of soy isoflavones in postmenopausal women with breast cancer, we are of the opinion that isoflavones should not be indicated because the studies are controversial.

With respect to ovarian cancer estrogen dependent, Lee et al. (2014) in a case-control study conducted in southern China have referred protective effect in women who consume at least 120 g / day of isoflavones compared to those consuming less than 61 g/day. As a justification, they believe that this fact is related to the induction of apoptosis and inhibition of the growth and proliferation of tumor cells (Chen and Anderson, 2001; Gercel-Taylor et al., 2004; Gossner et al., 2007). They further suggest that isoflavones may stimulate the production of liver globulins which bind to sex hormone binding globulin (SHBG), and so leads to a reduction of bioavailable estrogen in plasma (Adlercreutz et al., 1987). Another possible mechanism is the inhibition of the activity of ovarian aromatase, an enzyme that converts androgens to estrogens, as shown in vitro (Pelissero et al., 1996).

Regarding other symptoms, meta-analysis of randomized controlled trials to evaluate the efficacy in vasomotor symptoms and the side effects of phytoestrogens in postmenopausal women showed that there is much disagreement among authors. However, in the short run, most of them referred improvement in relief of these symptoms, as well as better quality of life and non-occurrence of side effects (Chen et al., 2014; Messina, 2014; Thomas et al., 2014).

With respect to the genitals during menopause, atrophy ranges from 10% to 50%, because the lack of circulating estrogen promotes the decrease of collagen and elasticity, which leads to the vulvovaginal atrophy and dryness by decreased lubrication. This causes discomfort and dyspareunia (pain) during sexual intercourse (Levine et al., 2008).

Nevertheless, epidemiological studies of postmenopausal women under the isoflavones gel describe an improving of the vaginal trophism, which consequently led to amelioration in vaginal symptoms, pH, and increased expression of estrogen receptors, all indicators to that isoflavones may be a good therapeutic option for the relief of vulvovaginal atrophy (Lima et al., 2013).

Experimental studies with rats treated with isoflavones for a long period showed trophism of the vaginal epithelium, thereby confirming the trophic effect due to the use of isoflavones (Carbonel et al., 2011).

In addition to the listed shares, isoflavones are also known for their antioxidant properties, including the ability to regulate the expression and enzymatic activity of the antioxidant system; they inhibit the oxidation of cellular components by direct sequestration of free radicals via phenolic rings or by its ability to chelate metal ions involved in oxidative process (Bertoncini et al., 2010; Erba et al., 2012).

Excessive production of free radicals can create an unsuitable environment for normal physiological reactions, leading to a series of reproductive system diseases, such as endometriosis, polycystic ovary syndrome (PCOS) and unexplained infertility or sterility without apparent cause (Sharma and Agarwal, 2004; Banerje et al., 2013).

In the female reproductive system, free radicals play an important role in the regulation of several signaling pathways in folliculogenesis, oocyte maturation, cyclical endometrial changes, and implantation of the embryo. Thus, oxidative stress influences the entire reproductive life of the woman, including the modulation of the decline of fertility regarding age (De Bruin et al., 2002).

Study carried out in infertile women showed that consumption of soy in the diet during treatment with assisted reproductive technology seems to be positive about to the probability of pregnancy (Vanegas et al., 2014). Likewise, Unfer et al. (2004) observed that the rate of pregnancy/childbirth ongoing among women undergoing in vitro fertilization and receiving isoflavone supplementation was nearly twice than to the women who did not use soflavones. Another study in Japanese infertile women showed that diets rich in daidzein and genistein can reduce the risk of deep endometriosis (Tsuchiya et al., 2007).

The reduction of oxidative stress by soy isoflavones has been demonstrated in several vivo models (Jungbauer and Medjakovic, 2014). Genistein and daidzein have also been associated with decreased risk of chronic diseases, such as neurodegenerative cardiovascular, metabolic and cancer diseases, due in part to the antioxidant activity of the isoflavones (Yoon and Park, 2014).

An experimental study with sexually mature rats showed that treatment with genistein reduces the follicular atresia and increased the number of surviving ovarian follicles, suggesting their benefit in prolonging the reproductive life (Zhuang et al., 2010).

CONCLUSION

The findings about soy isoflavones as antioxidants have expanded their usage possibilities, not only for application in the disorders resulting from menopause, but also to prevent and treat disturbances associated with increased oxidative stress. Double-blind randomized study in postmenopausal women who consumed soy isoflavones showed that the use for a long period does not affect the endometrial thickness as well as does not increase the hyperplasia and endometrial cancer rates (Quaas et al., 2013).

Studies performed in rats showed that genistein and daidzein induce various genomic responses in the uterus. In near normal doses, they do not stimulate proliferation and may be considered both agonists and selective estrogen receptor modulators (SERM) (Gaete et al.,

2012; Carbonel et al., 2014, Teixeira et al., 2014). Nevertheless, trophic effect of isoflavones on the endometrium when they were administered at high doses. Even so, when in combination with estrogens, isoflavones showed no additive effect (Francisco et al., 2013).

References

Adlercreutz, H., Heinonen, S. M. and Penalvo-Garcia, J. Phytoestrogens, cancer and coronary heart disease. *Biofactors*. 2004;22:229-36.

Adlercreutz, H., Höckerstedt, K., Bannwart, C., Bloigu, S., Hämäläinen, E., Fotsis, T., and Ollus, A. Effect of dietary components, including lignans and phytoestrogens, on enterohepatic circulation and liver metabolism of estrogens and on sex hormone binding globulin (SHBG). *J. Steroid Biochem*. 1987;27:1135-44.

Akiyama, T., Ishida, J., Nakagawa, S., Ogawara, H., Watanabe, S., Itoh, N., Shibuya, M., and Fukami, Y. Genistein, a specific inhibitor of tyrosine-specific protein kinases. *J. Biol. Chem*. 1987;262:5592-5.

Aso, T., Uchiyama, S., Matsumura, Y., Taguchi, M., Nozaki, M., Takamatsu, K., Ishizuka, B., Kubota, T., Mizunuma, H., and Ohta, H. A natural S-equol supplement alleviates hot flushes and other menopausal symptoms in equol nonproducing postmenopausal Japanese women. *J. Womens Health (Larchmt)*. 2012;21(1):92-100.

Avis, N. E., Stellato, R., Crawford, S., Bromberger, J., Ganz, P., Cain, V., and Kagawa-Singer, M. Is there a menopausal syndrome? Menopausal status and symptoms across racial/ethnic groups. *Soc. Sci. Med*. 2001;52(3):345-56.

Baechler, S. A., Schroeter, A., Walker, J., Aichinger, G., and Marko, D. Oxidative metabolism enhances the cytotoxic and genotoxic properties of the soy isoflavone daidzein. *Mol. Nutr. Food Res*. 2014;58(6):1269-81.

Banerjee, J., Shaeib, F., Maitra, D., Saed, G. M., Dai, J., Diamond, M. P., and Abu-Soud, H. M. Peroxynitrite affects the cumulus cell defense of metaphase II mouse oocytes leading to disruption of the spindle structure in vitro. *Fertil. Steril*. 2013;100(2):578-84.

Banerjee, S., Li, Y., Wang, Z., and Sarkar, F. H. Multi-targeted therapy of cancer by genistein. *Cancer Lett*. 2008;269(2):226-42.

Barnes, S., Peterson, G., Grubbs, C., and Setchell, K. Potential role of dietary isoflavones in the prevention of cancer. *Adv. Exp. Med. Biol*. 1994;354:135-47.

Bedell, S., Nachtigall, M. and Naftolin, F. The pros and cons of plant estrogens for menopause. *J. Steroid Biochem. Mol. Biol*. 2014;139:225-36.

Bertoncini, C. R., Meneghini, R. and Tolentino, H. Nearest-neighbor nitrogen and oxygen distances in the iron(II)-DNA complex studied by extended X-ray absorption fine structure. *Spectrochim. Acta A Mol. Biomol. Spectrosc*. 2010;77(4):908-10.

Carbonel, A. A., Calió, M. L., Santos, M. A., Bertoncini, C. R., Sasso, G. D., Simões, R. S., Simões, M J., and Soares Jr, J. M. Soybean isoflavones attenuate the expression of genes related to endometrial cancer risk. *Climacteric*. 2014;27:1-10.

Carbonel, A. A., Simões, R. S., Santos, R. H., Baracat, M. C., Simões, Mde J., Baracat, E. C., and Soares Jr, J. M. Effects of high-dose isoflavones on rat uterus. *Rev. Assoc. Med. Bras*. 2011;57(5):534-9.

Cavallini, A., Dinaro, E., Giocolano, A., Caringella, A. M., Ferreri, R., Tutino, V., and Loverro, G. Estrogen receptor (ER) and ER-related receptor expression in normal and atrophic human vagina. *Maturitas*. 2008;59(3):219-25.

Cederroth, C. R. and Nef, S. Soy, phytoestrogens and metabolism: A review. *Mol. Cell. Endocrinol*. 2009;304(1-2):30-42.

Chen, J., Lin, C., Yong, W., Ye, Y., and Huang, Z. Calycosin and genistein induce apoptosis by inactivation of HOTAIR/p-Akt signaling pathway in human breast cancer MCF-7 cells. *Cell. Physiol. Biochem*. 2015;35(2):722-8.

Chen, J., Zeng, J., Xin, M., Huang, W., and Chen, X. Formononetininduces cell cycle arrest of human breast cancer cells via IGF1/PI3K/Akt pathways *in vitro* and *in vivo*. *Horm. Metab. Res*. 2011;43:681-6.

Chen, M. N., Lin, C. C. and Liu, C. F. Efficacy of phytoestrogens for menopausal symptoms: a meta-analysis and systematic review. *Climacteric* 2014;17:1-10.

Chen, X. and Anderson, J. J. Isoflavones inhibit proliferation of ovarian cancer cells in vitro via an estrogen receptor-dependent pathway. *Nutr. Cancer*. 2001;41:165-71.

Choi, E. J. and Kim, G. H. Daidzein causes cell cycle arrest at the G1 and G2/M phases in human breast cancer MCF-7 and MDA-MB-453 cells. *Phytomedicine* 2008; 15(9):683-90.

Choi, Y. H., Lee, W. H., Park, K. Y., and Zhang, L. p53-independent induction of p21 (WAF1/CIP1), reduction of cyclin B1 and G2/M arrest by the isoflavone genistein in human prostate carcinoma cells. *Jpn. J. Cancer Res*. 2000;91(2):164-7.

De Bruin, J. P., Dorland, M., Spek, E. R., Posthuma, G., van Haaften, M., Looman, C. W., and te Velde, E. R. Ultrastructure of the resting ovarian follicle pool in healthy young women. *Biol. Reprod*. 2002;66:1151-60.

Di Virgilio, A. L., Iwami, K., Wätjen, W., Kahl, R., and Degen, G. H. Genotoxicity of the isoflavones genistein, daidzein and equol in V79 cells. *Toxicol. Lett*. 2004;151(1):151-62.

Dixon, R. A. and Ferreira, D. Genistein. *Phytochemistry*. 2002;60(3):205-11.

Erba, D., Casiraghi, M. C., Martinez-Conesa, C., Goi, G., and Massaccesi, L. Isoflavone supplementation reduces DNA oxidative damage and increases O-β-N-acetyl-D-glucosaminidase activity in healthy women. *Nutr. Res*. 2012;32:233-40.

Fitzpatrick, L. A. Soy isoflavones: hope or hype? *Maturitas*. 2003;44(Suppl. 1):S21-9.

Francisco, A. M., Carbonel, A. F., Simões, R. S., Soares Jr, J. M., Baracat, E. C., and Haidar, M. A. Do extracts of oral soybean augment the trophic effect of estrogen on the rat uterus? *Climacteric* 2013;16(1):161-8.

Franke, A. A., Lai, J. F. and Halm, B. M. Absorption, distribution, metabolism, and excretion of isoflavonoids after soy intake. *Arch. Biochem. Biophys*. 2014;559:24-8.

Gaete, L., Tchernitchin, A. N., Bustamante, R., Villena, J., Lemus, I., Gidekel, M., Cabrera, G., and Astorga, P. Daidzein-estrogen interaction in the rat uterus and its effect on human breast cancer cell growth. *J. Med. Food*. 2012;15(12):1081-90.

Gaete, L., Tchernitchin, A. N., Bustamante, R., Villena, J., Lemus, I., Gidekel, M., Cabrera, G., and Carrillo, O. Genistein selectively inhibits estrogen-induced cell proliferation and other responses to hormone stimulation in the prepubertal rat uterus. *J. Med. Food*. 2011; 14(12):1597-603.

Gercel-Taylor, C., Feitelson, A. K. and Taylor, D. D. Inhibitory effect of genistein and daidzein on ovarian cancer cell growth. *Anticancer Res*. 2004;24:795-800.

Gossner, G., Choi, M., Tan, L., Fogoros, S., Griffith, K. A., Kuenker, M., and Liu, J. R. Genistein-induced apoptosis and autophagocytosis in ovarian cancer cells. *Gynecol. Oncol.* 2007;105:23-30.

Gu, L., House, S. E., Prior, R. L., Fang, N., Ronis, M. J., Clarkson, T. B., Wilson, M. E., and Badger, T. M. Metabolic phenotype of isoflavones differ among female rats, pigs, monkeys, and women. *J. Nutr*. 2006;136(5):1215-21.

Guo, J. M., Xiao, B. X., Dai, D. J., Liu, Q., and Ma, H. H. Effects of daidzein on estrogen-receptor-positive and negative pancreatic cancer cells *in vitro*. *World J. Gastroenterol.* 2004;10(6):860-3.

Hewitt, A. L. and Singletary, K. W. Soy extract inhibits mammary adenocarcinoma growth in a syngeneic mouse model. *Cancer Lett.* 2003;192(2):133-43.

Hsieh, C. Y., Santell, R. C., Haslam, S. Z., and Helferich, W. G. Estrogenic effect of genistein on the growth of estrogen receptor-positive human breast cancer (MCF-7) cells in vitro and in vivo. *Cancer Res*. 1998;58:3833-8.

Hsu, A., Bray, T. M., Helferich, W. G., Doerge, D. R., and Ho, E. Differential effects of whole soy extract and soy isoflavones on apoptosis in prostate cancer cells. *Exp. Biol. Med.* 2010;235(1):90-7.

Hwang, W., Hou, Y. and Yao, W. Equol: a metabolite of soy isoflavones by intestinal microflora--a review. *Wei Sheng Wu Xue Bao.* 2013;53(12):1251-7.

Izumi, T., Piskula, M. K., Osawa, S., Obata, A., Tobe, K., Saito, M., Kataoka, S., Kubota, Y., and Kikuchi, M. Soy isoflavone aglycones are absorbed faster and in higher amounts than their glucosides in humans. *J. Nutr*. 2000;130(7):1695-9.

Jenks, B. H., Iwashita, S., Nakagawa, Y., Ragland, K., Lee, J., Carson, W. H., Ueno, T., and Uchiyama, S. A pilot study on the effects of S-equol compared to soy isoflavones on menopausal hot flash frequency. *J. Womens Health (Larchmt).* 2012;21(6):674-82.

Jian, L. Soy, isoflavones, and prostate cancer. *Mol. Nutr. Food Res*. 2009;53(2):217-26.

Jin, S., Zhang, Q. Y., Kang, X. M., Wang, J. X., and Zhao, W. H. Daidzein induces MCF-7 breast cancer cell apoptosis via the mitochondrial pathway. *Ann. Oncol.* 2010;21(2):263-8.

Jordan, V. C. Chemoprevention of breast cancer with selective oestrogen-receptor modulators. *Nat. Rev. Cancer*. 2007;7:46-53.

Ju, Y. H., Fultz, J., Allred, K. F., Doerge, D. R., and Helferich, W. G. Effects of dietary daidzein and its metabolite, equol, at physiological concentrations on the growth of estrogen-dependent human breast cancer (MCF-7) tumors implanted in ovariectomized athymic mice. *Carcinogenesis*. 2006;27(4):856-63.

Jungbauer, A. and Medjakovic, S. Phytoestrogens and the metabolic syndrome. *J. Steroid Biochem. Mol. Biol.* 2014;139:277-89.

Kajta, M., Rzemieniec, J., Litwa, E., Lason, W., Lenartowicz, M., Krzeptowski, W., and Wojtowicz, A. K. The key involvement of estrogen receptor β and G-protein-coupled receptor 30 in the neuroprotective action of daidzein. *Neuroscience*. 2013;238:345-60.

Khaodhiar, L., Ricciotti, H. A., Li, L., Pan, W., Schickel, M., Zhou, J., and Blackburn, G. L. Daidzein-rich isoflavoneaglycones are potentially effective in reducing hot flashes in menopausal women. *Menopause*. 2008;15(1):125-32.

Klein, C. B and King, A. A. Genistein genotoxicity: critical considerations of in vitro exposure dose. *Toxicol. Appl. Pharmacol*. 2007;224(1):1-11.

Ko, K. P. Isoflavones: chemistry, analysis, functions and effects on health and cancer. *Asian Pac. J. Cancer Prev*. 2014;15(17):7001-10.

Lee, A. H., Su, D., Pasalich, M., Tang, L., Binns, C. W., and Qiu, L. Soy and isoflavone intake associated with reduced risk of ovarian cancer in southern Chinese women. *Nutr. Res*. 2014;34(4):302-7.

Levine, K. B., Williams, R. E. and Hartmann, K. E. Vulvovaginal atrophy is strongly associated with female sexual dysfunction among sexually active postmenopausal women. *Menopause*. 2008;15(4 Pt 1):661-6.

Li, H., Yu, Y. Y., Hu, X., and Cao, S. W. Research on the interactions between genistein and its glucosides with DNA. *Guang Pu Xue Yu Guang Pu Fen Xi*. 2008;28(8):1905-9.

Li, T., Zhao, X., Mo, Z., Huang, W., Yan, H., Ling, Z., and Ye, Y. Formononetin promotes cell cycle arrest via down regulation of Akt/Cyclin D1/CDK4 in human prostate cancer cells. *Cell. Physiol. Biochem*. 2014;34(4):1351-8.

Li, Y. and Anand-Srivastava, M. B. Implication of multiple signaling pathways in the regulation of angiotensin II induced enhanced expression of Giα proteins in vascular smooth muscle cells. *Can. J. Physiol. Pharmacol*. 2012;90(8):1105-16.

Lima, S. M., Bernardo, B. F., Yamada, S. S., Reis, B. F., da Silva, G. M., and Galvão, M. A. Effects of Glycine max (L.) Merr. soy isoflavone vaginal gel on epithelium morphology and estrogen receptor expression in postmenopausal women: a 12-week, randomized, double-blind, placebo-controlled trial. *Maturitas*. 2014;78(3):205-11.

Lima, S. M., Yamada, S. S., Reis, B. F., Postigo, S., Galvão da Silva, M. A., and Aoki, T. Effective treatment of vaginal atrophy with isoflavone vaginal gel. *Maturitas*. 2013; 74 (3):252-8.

Liu, Z., Li, W., Sun, J., Liu, C., Zeng, Q., Huang, J., Yu, B., and Huo, J. Intake of soy foods and soy isoflavones by rural adult women in China. *Asia Pac. J. Clin. Nutr*. 2004;13(2): 204-9.

Luo, L. J., Liu, F., Lin, Z. K., Xie, Y. F., Xu, J. L., Tong, Q. C., and Shu, R. Genistein regulates the IL-1 beta induced activation of MAPKs in human periodontal ligament cells through G protein-coupled receptor 30. *Arch. Biochem. Biophys*. 2012;522(1):9-16.

Mai, Z., Blackburn, G. L. and Zhou, J. R. Soy phytochemicals synergistically enhance the preventive effect of tamoxifen on the growth of estrogen-dependent human breast carcinoma in mice. *Carcinogenesis*. 2007;28(6):1217-23.

Mateos-Aparicio, I., Redondo Cuenca, A., Villanueva-Suárez, M. J., and Zapata-Revilla, M. A. Soybean, a promising health source. *Nutr. Hosp*. 2008;23(4):305-12.

Messina, M. Soybean isoflavones warrant greater consideration as a treatment for the alleviation of menopausal hot flashes. *Womens Health* (Lond. Engl.). 2014;10(6):549-53.

Moiseeva, E. P., Almeida, G. M., Jones, G. D., and Manson, M. M. Extended treatment with physiologic concentrations of dietary phytochemicals results in altered gene expression, reduced growth, and apoptosis of cancer cells. *Mol. Cancer Ther*. 2007;6(11):3071-9.

Morimoto, Y., Maskarinec, G., Park, S. Y., Ettienne, R., Matsuno, R. K., Long, C., Steffen, A. D., Henderson, B. E., Kolonel, L. N., Le Marchand, L., and Wilkens, L. R. Dietary isoflavone intake is not statistically significantly associated with breast cancer risk in the Multiethnic Cohort. *Br. J. Nutr*. 2014;112(6):976-83.

Morito, K., Hirose, T., Kinjo, J., Hirakawa, T., Okawa, M., Nohara, T., Ogawa, S., Inoue, S., Muramatsu, M., and Masamune, Y. Interaction of phytoestrogens with estrogen receptorsalpha and beta. *Biol. Pharm. Bull*. 2001;24(4):351-6.

Morton, M. S., Arisaka, O., Miyake, N., Morgan, L. D., and Evans, B. A. Phytoestrogen concentrations in serum from Japanese men and women over forty years of age. *J. Nutr.* 2002;132(10):3168-71.

Oseni, T., Patel, R., Pyle, J., and Jordan, V. C. Selective estrogen receptor modulators and phytoestrogens. *Planta Med.* 2008;74(13):1656-65.

Patisaul, H. B. and Jefferson, W. The pros and cons of phytoestrogens. *Front. Neuroendocrinol.* 2010;31(4):400-19.

Pavese, J. M., Farmer, R. L. and Bergan, R. C. Inhibition of cancer cell invasion and metastasis by genistein. *Cancer Metastasis Rev.* 2010;29(3): 465-82.

Pelissero, C., Lenczowski, M. J., Chinzi, D., Davail-Cuisset, B., Sumpter, J. P., and Fostier, A. Effects of flavonoids on aromatase activity, an in vitro study. *J. Steroid Biochem. Mol. Biol.* 1996;57:215-23.

Pilšáková, L., Riečanský, I. and Jagla, F. The physiological actions of isoflavone phytoestrogens. *Physiol. Res.* 2010;59(5):651-64.

Prossnitz, E. R. and Barton, M. The G-protein-coupled estrogen receptor GPER in health and disease. *Nat. Rev. Endocrinol.* 2011;7(12):715-26.

Quaas, A. M., Kono, N., Mack, W. J., Hodis, H. N., Felix, J. C., Paulson, R. J., and Shoupe, D. Effect of isoflavone soy protein supplementation on endometrial thickness, hyperplasia, and endometrial cancer risk in postmenopausal women: a randomized controlled trial. *Menopause.* 2013;20(8):840-4.

Ricciotti, H. A., Khaodhiar, L. and Blackburn, G. L. Daidzein-rich isoflavone-aglyconesfor menopausal symptoms. *Int. J. Gynaecol. Obstet.* 2005;89(1):65-6.

Rossouw, J. E., Anderson, G. L., Prentice, R. L., LaCroix, A. Z., Kooperberg, C., Stefanick, M. L., Jackson, R. D., Beresford, S. A., Howard, B. V., Johnson, K. C., Kotchen, J. M., Ockene, J., and Writing Group for the Women's Health Initiative Investigators. Risks and benefits of estrogen plus progestin in healthy postmenopausal women: principal results From the Women's Health Initiative randomized controlled trial. *JAMA.* 2002;288(3): 321-33.

Setchell, K. D., Brown, N. M. and Lydeking-Olsen, E. The clinical importance of themetabolite equol-a clue to the effectiveness of soy and its isoflavones. *J. Nutr.* 2002; 132(12):3577-84.

Setchell, K. D., Brown, N. M., Desai, P. B., Zimmer-Nechimias, L., Wolfe, B., Jakate, A. S., Creutzinger, V., and Heubi, J. E. Bioavailability, disposition, and dose-response effects of soy isoflavones when consumed by healthy women at physiologically typical dietary intakes. *J. Nutr.* 2003;133(4):1027-35.

Setchell, K. D., Brzezinski, A., Brown, N. M., Desai, P. B., Melhem, M., Meredith, T., Zimmer-Nechimias, L., Wolfe, B., Cohen, Y., and Blatt, Y. Pharmacokinetics of a slow-release formulation of soybean isoflavones in healthy postmenopausal women. *J. Agric. Food Chem.* 2005;53(6):1938-44.

Sharma, R. K. and Agarwal, A. Role of reactive oxygen species in gynecologic diseases. *Reprod. Med. Bio.* 2004;3:177-99.

Shike, M., Doane, A. S., Russo, L., Cabal, R., Reis-Filo, J., Gerald, W., Cody, H., Khanin, R., Bromberg, J., and Norton, L. The effects of soy supplementation on gene expression inbreast cancer: a randomized placebo-controlled study. *J. Natl. Cancer Inst.* 2014;106 (9):189.

Smith, C. L. and O'Malley, B. W. Coregulator function: a key to understanding tissue specificity of selective receptor modulators. *Endocr. Rev.* 2004;25:45-71.

Somekawa, Y., Chiguchi, M., Ishibashi, T., and Aso, T. Soy intake related to menopausal symptoms, serum lipids, and bone mineral density in postmenopausal Japanese women. *Obstet. Gynecol.* 2001;97(1):109-15.

Teixeira, C. P., Simões, R. S., Santos, M. A., Calió, M. L., Soares Jr, J. M., Simões, M. J., Bertoncini, C. R., Higa, E. M., and Carbonel, A. F. Soybean concentrated extract counteracts oxidative stress in the uterus of rats. *Climacteric.* 2014;17:402-9.

Tepper, C. G., Vinall, R. L., Wee, C. B., Xue, L., Shi, X. B., Burich, R., Mack, P. C., and de Vere White, R. W. GCP-mediated growth inhibition and apoptosis of prostate cancer cells via androgen receptor-dependent and -independent mechanisms. *Prostate.* 2007;67 (5):521-35.

Thomas, A. J., Ismail, R., Taylor-Swanson, L., Cray, L., Schnall, J. G., Mitchell, E. S., and Woods, N. F. Effects of isoflavones and amino acid therapies for hot flashes and co-occurring symptoms during the menopausal transition and early postmenopause: a systematic review. *Maturitas.* 2014;78(4):263-76.

Touny, L. H. and Banerjee, P. P. Identification of both Myt-1 and Wee-1 as necessarymediators of the p21-independent inactivation of the cdc-2/cyclin B1 complex and growth inhibition of TRAMP cancer cells by genistein. *Prostate.* 2006;66(14):1542-55.

Tsuchiya, M., Miura, T., Hanaoka, T., Iwasaki, M., Sasaki, H., Tanaka, T., Nakao, H., Katoh, T., Ikenoue, T., Kabuto, M., and Tsugane, S. Effect of soy isoflavones on endometriosis: interaction with estrogen receptor 2 gene polymorphism. *Epidemiology.* 2007;18(3):402-8.

Unfer, V., Casini, M. L., Gerli, S., Costabile, L., Mignosa, M., and diRenzo, G. C. Phytoestrogens may improve the pregnancy rate in in vitro fertilization–embryo transfer cycles: a prospective, controlled, randomized trial. *Fertil. Steril.* 2004;82:1509-13.

Usui, T. Pharmaceutical prospects of phytoestrogens. *Endocr. J.* 2006;53(1):7-20.

Vanegas, J. C., Afeiche, M. C., Gaskins, A. J., Mínguez-Alarcón, L., Williams, P. L., Wright, D. L., Toth, T. L., Hauser, R., and Chavarro, J. E. Soy food intake and treatment outcomes of women undergoing assisted reproductive technology. *Fertil. Steril.* 2015;pii: S0015-0282(14)02529-1. doi: 10.1016/j.fertnstert.2014.12.104. [Epub. ahead of print].

Watanabe, S., Haba, R., Terashima, K., Arai, Y., Miura, T., Chiba, H., Takamatsu, K. Antioxidant activity of soya hypocotyl tea in humans. *Biofactors.* 2000;12(1-4):227-32.

Wegrzyn, G., Jakóbkiewicz-Banecka, J., Gabig-Cimińska, M., Piotrowska, E., Narajczyk, M., Kloska, A., Malinowska, M., Dziedzic, D., Gołebiewska, I., Moskot, M., and Wegrzyn, A. Genistein: a natural isoflavone with a potential for treatment of genetic diseases. *Biochem. Soc. Trans.* 2010;38(2):695-701.

Wu, A. H., Koh, W. P., Wang, R., Lee, H. P., and Yu, M. C. Soy intake and breast cancer risk in Sinapore Chinese Health Study. *Br. J. Cancer* 2008; 99:196-200.

Wu, A. H., Lee, E. and Vigen, C. Soy isoflavones and breast cancer. *Am. Soc. Clin. Oncol. Educ. Book.* 2013;2013:102-6.

Xu, L., Du, B. and Xu, B. A systematic, comparative study on the beneficial health components and antioxidant activities of commercially fermented soy products marketed in China. *Food Chem.* 2015;174:202-13.

Yoon, G. A. and Park, S. Antioxidant action of soy isoflavones on oxidative stress and antioxidant enzyme activities in exercised rats. *Nutr. Res. Pract.* 2014;8(6):618-24.

Yu, X., Zhu, J., Mi, M., Chen, W., Pan, Q., and Wei, M. Anti-angiogenic genistein inhibits VEGF-induced endothelial cell activation by decreasing PTK activity and MAPK activation. *Med. Oncol.* 2012;29(1):349-57.

Zhao, E. and Mu, Q. Phytoestrogen biological actions on mammalian reproductive system and cancer growth. *Sci. Pharm.* 2011;79(1):1-20.

Zheng, W., Hou, Y. and Yao, W. Equol: a metabolite of soy isoflavones by intestinal microflora-a review. *Wei Sheng Wu Xue Bao*. 2013;53(12):1251-7.

Zhuang, X. L., Fu, Y. C., Xu, J. J., Kong, X. X., Chen, Z. G., and Luo, L. L. Effects of genistein on ovarian follicular development and ovarian life span in rats. *Fitoterapia.* 2010;81(8):998-1102.

Zubik, L. and Meydani, M. Bioavailability of soybean isoflavones from aglycone and glucoside forms in American women. *Am. J. Clin. Nutr.* 2003;77(6):1459-65.

In: Genistein and Daidzein
Editor: Masayoshi Yamaguchi

ISBN: 978-1-63483-194-9

Chapter 2

DAIDZEIN AND GENISTEIN: ENDOTHELIAL ACTION OF AN AGLYCONE-RICH EXTRACT

Camila Marques de Andrade[1,*], Maria José, Vieira Fonseca[2] and Maria Regina Torqueti Toloi[1]

[1]University of São Paulo, Faculty of Pharmaceutical Sciences of Ribeirão Preto, Department of Clinical, Toxicological and Bromatological Analysis, Ribeirão Preto, Brazil

[2]University of São Paulo, Faculty of Pharmaceutical Sciences of Ribeirão Preto, Department of Pharmaceutical Sciences, Ribeirão Preto, Brazil

ABSTRACT

Oestrogen replacement therapy is not thought to be a safe treatment for prevention of cardiovascular disease in menopausal women; isoflavones are a possible alternative. The early stage of atherosclerosis is characterized by increased levels of inflammatory cytokines, oxidative stress, and the expression of adhesion molecules (CAMs) on vascular endothelium. The endothelium produces nitric oxide (NO), a vasoprotective and anti-atherosclerotic agent. Oestrogen has beneficial effects on the cardiovascular system by enhancing NO production. Oestrogen-like compounds such as isoflavones are also suggested increase NO production, exerting anti-atherogenic effects. Therefore, compounds that suppress CAMs, reduce oxidative stress or increase NO may be usable as therapeutic agents against the development of atherosclerosis in menopausal women. Soy isoflavones exert anti-atherogenic effects. Isoflavones are present in soy foods mainly as glucosides. Soy isoflavone aglycones, the biological active estrogenlike compounds, are absorbed faster and in higher amounts than their glucoside derivatives. In addition, the aglycone forms have a higher biological activity implying that isoflavone aglycone-rich

* Corresponding author: Camila Marques de Andrade. University of São Paulo, Faculty of Pharmaceutical Sciences of Ribeirão Preto, Department of Clinical, Toxicological and Bromatological Analysis, Ribeirão Preto, Brazil. E-mail: a_m.andrade@hotmail.com.

products may be more effective than the ones rich in glycosides in preventing chronic diseases such as coronary heart disease.

Objective: In this chapter we review and compiled all the results already published and present some unpublished results about our study comparing the effects of an extract of soybean fermented by *Aspergillus awamori* (a β-glucosidases-producing filamentous fungi) on which polyphenol glucosides were bioconverted to aglycone forms (BSE) with a non-fermented extract (SE) on adhesion molecules expression (VCAM-1, ICAM-1 and E-selectin), oxidative stress induced by hydrogen peroxide and on production of nitric oxide, prostaglandin E2 and Endothelin-1 *in vitro* in human endothelial cells (HUVEC).

Results: Bioconverted soybean extract (BSE) decreased VCAM-1, ICAM-1 and E-selectin on cell surface and on cell culture supernatant, decreased the oxidative stress induced by hydrogen peroxide and enhanced endothelin-1, nitric oxide and prostaglandin E2 production, while the unfermented extract also decreased adhesion molecules expression and oxidative stress; enhanced endothelin-1 production but was not able to increase nitric oxide and prostaglandin E2 production.

Conclusion: The aglycone -rich soybean extract was able to reduce adhesion molecules expression and oxidative stress, being more effective to reduce sICAM-1 expression and only the aglycone-rich forms soybean extracts was able to increase nitric oxide and prostaglandin E2 production, demonstrating that in edothelial cells *in vitro*, may be usable as therapeutic agents against the development of atherosclerosis.

Keywords: Phytoestrogens, adhesion molecules, nitric oxide, endothelin-1, prostaglandin E2, oxidative stress

Introduction

Cardiovascular diseases (CVD) are associated with oxidative stress and endothelial dysfunction. CVDs are the major cause of death among menopausal women in developed countries (Maturana et al., 2007), which may be attributed to the loss of cardiovascular protection through the deficiency of estrogen that occurs in this period (Mendelsohn; Karas, 2005).

Observational studies in postmenopausal women suggested that estrogen replacement therapy (HRT) reduces the risk of cardiovascular disease (Barrett-Connor; Bush, 1991). One of the main reasons that HRT reduces the risk of CVD development is its favorable impact on the vascular endothelium which promotes the synthesis of nitric oxide (NO) and other vasoactive factors (Chambliss; Shaul, 2002).

However, two large studies, the HERS and WHI, questioned the positive effects of HRT and warned of adverse effects: increased incidence of breast and endometrial cancer in long-term treatment (Hurley et al., 1998).

Since hormone replacement therapy was contradicted by these studies, alternatives therapies are being sought. Studies have indicated the effects of phytoestrogens (PE), non-steroidal compounds that promote estrogenic effects in mammals and are structurally similar to 17β-estradiol (E2). The various biological activities are due in part to the ability of these compounds to act as agonists in the estrogen receptors (ERs), promoting estrogenic effects, or as antagonists, blocking or changing the ERs and thus preventing against estrogenic activity (Kuiper et al., 1998).

Among the PE, isoflavones stand out, being primarily found in the *Glycine max* soybean. The current interest in these compounds comes from the observation that populations with high soy consumption (Asian populations) have a lower incidence of cardiovascular disease and lower incidence of complaints of menopausal symptoms, suggesting a direct relationship between soy consumption and improved quality of life during menopause (Quella et al., 2000). Chemical differences in soy isoflavones may result in different biological activities. Aglycones of soy isoflavones are absorbed more quickly and in higher doses than their glycosidic derivatives in humans (Chien et al., 2006).

Moreover, they present high biological activity (Lin et al., 2006) and, as a result, aglycone-rich products of isoflavones may be more effective than their glycosidic forms in the prevention of chronic diseases such as coronary diseases (Lee et al., 2005). A soybean extract biotransformed by the fungus *Aspergillus awamori* has been produced in the Faculty of Pharmaceutical Sciences of Ribeirao Preto – USP-Brazil. This extract is rich in aglycone forms of genistein and daidzein (Georgetti et al., 2009), and may have a greater estrogenic potential in face of pro-inflammatory factors and endothelial-derived factors (De Andrade et al., 2014A), (De Andrade et al., 2014B).

Vascular Endothelial

The vascular endothelium is a single layer of squamous and elongated cells, which covers the luminal surface of all blood vessels. Endothelial cells (EC) form the lumen of the vessels and are settled in the subendothelial area formed by the extracellular matrix, which extends through the internal elastic lamina. Participates in the regulation of vascular tone, the adherence of monocytes in thrombogenesis, inflammation, lipid metabolism and remodeling of vessel growth (Endermann, Schiffrin, 2004).

The endothelial tissue can inhibit the proliferation of vascular smooth muscle, thus preventing migration of vascular smooth muscle cells by direct mechanisms. It also plays crucial role in vascular homeostasis and in prevention of the onset and progress of cardiovascular disease, through the synthesis of factors working in smooth muscle cells, thus acting in the modulation of vascular tone, through the balance of the release of vasoactive substances (Mahmoudi; Curzen; Gallacher, 2007).

Endothelium-derived factors that promote vasodilation, (EDVF Endothelium-derived vasodilator factors) are: nitric oxide (NO - Nitric Oxide), the endothelium-derived hyperpolarizing factor (EDHF - Endothelium- Hyperpolarizing Derived Factor) prostacyclin (PGI2 - Prostaciclyn I2); the factors that promote vasoconstriction are endothelin-1 (ET-1 - Endothelin-1) and prostanoids prostaglandin H2 (PGH2 - prostaglandine H2), thromboxane A2 (TXA2 - Thromboxane A2) and angiotensin. Of all the factors that act on the smooth muscle highlight NO as the main mediator of vascular tone (Stankevicius et al., 2003).

Oxidative Stress

Oxidative stress is the term used to describe the imbalance in the balance between oxidants and antioxidants (Birben et al., 2012) with the presence of reactive oxygen species (ROS) exceeding neutralization capacity of antioxidants.

These reactive oxygen species can damage proteins, lipids and DNA, altering the structure and function of the body (Roberts and Hubel, 2004).

Reactive oxygen species (ROS) are produced from oxygen as a result of normal cellular metabolism. It can be divided into two groups, free radicals and non-radical. Molecules containing one or more unpaired electrons become unstable and tend to react with other molecules to pair their electrons and generate more stable species (Gueteens et al., 2002).

By providing the reactive molecules are called free radicals. When two free radicals divide their unpaired electrons, non-radical forms are created. The three main reactive species of oxygen physiologically significant are superoxide anion (O2-), hydroxyl radical (OH) and hydrogen peroxide (H2O2) (Birben et al., 2012).

The oxidative stress may be from endogenous sources or exogenous sources (Ray et al., 2001). The main source of endogenous ROS is aerobic respiration, but they are also produced by peroxisomal β-oxidation of fatty acids, microsomal cytochrome P450, metabolism of xenobiotic compounds, stimulation of phagocytosis by pathogens and lipopolysaccharide, arginine metabolism and tissue-specific enzymes (Nicholls and Budd 2000).

Oxidative stress plays an important role in inflammation and contributes in a variety of degenerative diseases such as atherosclerosis and cardiovascular diseases (Galli et al., 2005), it causes tissue damage or even cell death by two mechanisms: necrosis and apoptosis (Gueteens, et al., 2002).

Cells present antioxidant defenses against reactive oxygen species, including detoxifying enzymes and scavengers of exogenous free radicals (vitamins). The main enzymes that convert ROS in less reactive molecules are peroxide dismutase, catalase and glutathione peroxidase (Zhu et al., 2008).

In healthy individuals, antioxidants are the first line of defense against reactive oxygen species, removing them before they cause damage to molecules and preventing spread through the interruption of chain reaction of free radicals.

Antioxidants donate an electron to the free radical (Wang et al., 2008). Fruits and vegetables are major dietary antioxidants being particularly rich in vitamins (A, C and E), trace elements and polyphenols (Blockhina et al., 2003).

Endothelium-Derived Factors

Nitric Oxide (NO)

In 1980, a study by Furchgott and Zawadsky, established the importance of the endothelium in the control of vascular tone. In this study the researchers demonstrated that the vasodilator effect of acetylcholine depended on the release of an endothelium-derived relaxing factor. Years later, in 1987, Palmer and co-workers demonstrated that aortic EC cultures did release NO in concentrations sufficient to promote vasodilation, indicating that the endothelial derived relaxing factor was NO (Stankevicius et al., 2003).

Under physiological conditions, the vascular relaxation occurs when the EC membrane receptors are activated by acetylcholine, bradykinin, adenosine diphosphate and other, stimulating the formation of calcium calmodulin complex leading to activation of the enzyme endothelial nitric oxide synthase (eNOS) present in such cells and the production of NO. The eNOS is anchored to the endothelial cell membrane, which promotes the release of NO near the muscle layer of the vessel (Busconi; Michael, 1994).

NO is also a potent inhibitor of platelet aggregation which, by leaving the endothelial cell toward the bloodstream, can penetrate into platelets, especially in which are juxtaposed with the vessel wall. Within the platelets, in a manner analogous to that discussed for the muscle cells, NO promotes an increase in cGMP and consequent reduction of free Ca ++. Since the Ca ++ is essential for platelet aggregation, this process will also be inhibited (Wolin, 1996).

Prostaglandins (PGE)

PGE are fatty acids of 20 carbon atoms having in its basic structure the prostanoic acid, which have 20 carbons with a cyclopentane ring and two side chains.

Also have a hydroxyl group at position 15 and a double bond between carbons 13:14 at the trans position. PGE is synthesized from a primary precursor, the arachidonic acid. This acid is released from phospholipids of cell membranes by the action of phospholipase A2 by a process controlled by hormones and other stimuli.

There are two main routes to PGE synthesis from arachidonic acid: the path of lipoxygenase and cyclooxygenase route (COX) (Parkington; Colerman; Tare, 2004). Since the discovery of PGE the elucidation of its role in hemostasis and in cardiovascular diseases is becoming increasingly important, especially the so-called balance between TXA 2 and PGE. Under normal conditions, when stimulated by COX2 (cyclooxygenase 2) the endothelial cells form PGE, which is antithrombogenic and has a vasodilator effect by inhibiting platelet adhesion and aggregation.

Endothelin-1

Endothelin-1 (ET-1) was first isolated in EC culture pigs; constitute a family of peptides consisting of 21 amino acids. The endothelin family has three members: endothelin-1 (ET-1), endothelin-2 (ET-2) and endothelin-3 (ET-3) (Marasciulo; Montagnani; Potenza, 2006). ET-1 is produced in response to physical stimuli such as stress, hypoxia, cytokines, and in response to the action of vasoactive hormones among them: epinephrine and angiotensin II. They are mainly synthesized by EC, but also by other as macrophages and vascular smooth muscle cells (Webb; Haynes, 1995).

In the vascular system, ET-1 acts as a potent vasoconstrictor and has an important physiological role in the regulation of vascular tone, stimulating the generation of vascular mediators such as NO, PGE and platelet activating factors (Haller, 1997). The NO and ET-1 are the endothelium-derived factors most relevant in the case of cardiovascular diseases. NO causes vasodilation, inhibits platelet aggregation, inhibit ET-1 production, suppresses proliferation of muscle cells and acts as antithrombogenic factor. Alternatively, ET-1 acts in the opposite way to NO, causing potent vasoconstriction, increasing the adhesion of monocytes, macrophage activating and promoting the proliferation and migration of smooth muscle in the vessel (Schneider; Boesen; Pollock, 2007).

Adhesion Molecules and Inflammatory Process

The inflammatory process controls the complex response of vascularized tissue to cellular injury. A progressive sequence of paired connections between adhesion molecules expressed by endothelial cells and leukocytes binding causes transmigration of leukocytes to endothelium. These steps are described as adhesion, rolling, leukocyte activation and firm adhesion. The final step of this sequence, transendothelial migration, occurs at the junctions between endothelial cells (Liu et al., 2004; Muller, 2002).

Adhesion molecules belonging to four major protein families: selectins, selectin ligands, integrins and members of the immunoglobulin family. Selectins are transmembrane glycoproteins with homologous extracellular domains that vary in length and intracellular cytoplasmic domain without homology that induce "rolling" of leukocytes. Are molecules that bind to carbohydrates, forming three subfamilies (Huo; Ley, 2001): P -selectins (CD62P, 120-140kd) stored in Weibel-Palade bodies of endothelial cells and the alpha granules of platelets. Is expressed by endothelial cells and activated platelets, rapidly translocated to the surface of cells in response to a proinflammatory stimulation (Tedder et al., 1995). The glycoprotein-1 P-selectin (PSGL-1 - P-selectin glycoprotein ligand-1) is the main ligand for P-selectin and binding of these two factors is crucial for the "rolling" of leukocytes at the site of inflammation (Kansas, 1996). P-selectin reaches cell surface minutes after stimulation of the endothelial cell; E-selectin (CD62E, 115kd 107-), one subfamily of selectin is not expressed under normal physiological conditions, but is synthesized and exported to the surface of activated endothelial cells in inflammatory situations (Bevilacqua et al., 1987), and finally L-selectin (CD62L, 95kd) is expressed on all granulocytes, monocytes and in most lymphocytes (Ley, 2003).

Integrins and members of the immunoglobulin superfamily are also mediators of leukocyte adhesion and transmigration to vascular wall. Integrins consist of a common beta subunit, linked with multiple alpha subunits. Several extracellular IgG-like repeats domains, a transmembrane region and a short cytoplasmic domain, characterize the members of the immunoglobulin superfamily (Carlos; Harlan, 1994). Among them are the intercellular adhesion molecules ICAM 1, 2 and 3 (CD54, or ICAM - intercellular adhesion molecule) of 97-114 kd, constitutively expressed at low levels; adhesion molecules vascular cell-1, VCAM-1 (CD106 VCAM or - vascular cell adhesion molecule), 100-110 kD expressed by activated endothelial cells and follicular dendritic cells; and endothelial adhesion molecule-1 cell-platelet, or PECAM CD31 (PECAM - platelet endothelial cell adhesion molecule). The main function of these molecules is to act as integrin receptors. Inhibition of the interaction between adhesion molecules and their ligands reduces leukocyte adhesion, diapedesis and in some cases also affects the "rolling" of leukocytes preventing thus atherogenesis (Collins et al., 2000).

Hormone Replacement Therapy

The creation of the first synthetic estrogen (diethylstilbestrol) initiated in 1960 the use of these drugs by women. Initially as oral contraceptives, passing in 1966 to be used by menopausal women and called estrogen replacement therapy (ERT).

This hormone was used alone and in high doses. In 1975, a study showed that the risk of endometrial cancer increased 7.6 times in women using estrogen (Sellman, 1998).

From there this hormone has become associated with a synthetic progesterone, progestin, mimicking the natural action of progesterone to protect the uterus against the proliferative effects of estrogen. Then came the hormone replacement therapy (Sellman, 1998).

Its use was widespread and this therapy has become used to treat symptoms of menopause such as hot flashes and urogenital atrophy, for short periods.

Using estrogen alone, in various preparations and routes of administration or combined with progesterone (Rymmer; Morris, 2000), also began to be used in the prevention of chronic conditions such as osteoporosis (Humphrey; Chan; Sox, 2002), to increase bone mineral density, reducing the risk of fractures (Weiss; Ure; Ballard, 1980; Naessen; Persson; Adami, 1990), dementia and cognitive disorders. Alzheimer's disease (Henderson et al., 1994; Tang et al., 1996; Yaffe et al., 1998) and cardiovascular diseases (Grodstein et al., 1996).

During this period many studies were conducted with always controversial results, which attested from thromboembolic disorders (Cano, Van Baal, 2001) to increased risk of coronary heart disease (Grodstein et al., 1996). However, in 2002, the Women's Health Initiative Hormone Program, studied 16,608 women using HRT combining equine estrogen and medroxyprogesterone acetate or placebo, assessing the risk and benefits of using hormones at menopause. After five years of follow-up, the study was stopped because the risks: breast cancer, heart attacks, heart disease and thromboembolism, outweighed the benefits: reducing the number of fractures and colon and rectal cancer, concluding that the HRT regimen used not offered cardioprotective effect and is only recommended for relief of menopausal symptoms for short periods (WHI, 2002). Thus, a conservative position to start hormone replacement therapy long term came to be taken (Stork et al., 2002), and women have become hesitant to start or continue hormone treatment, becoming necessary to seek new therapeutic alternatives. In this context, many groups have begun to explore the estrogenic activity of natural compounds found in high concentrations in certain plants, the so-called phytoestrogens, starting from the observation that Asian women, which maintains high intake of dietary phytoestrogens have a lower incidence of menopausal symptoms and cardiovascular disease (Nagata, Takatsuka, Shimizu, 2002).

Phytoestrogens

Phytoestrogens are polyphenolic, nonsteroidal components of plant with similar biological activity of estrogen and bind to human estrogen receptor, exerting the characteristics of the endogenous hormone.

According to their chemical structure phytoestrogens can be classified into four groups: isoflavones, flavonoids, lignins and estilbenes (Wuttke et al., 2002). Isoflavones are the most studied group and were primarily found in soybean *Glycine max* and later discovered in other plants: Mexican Wild Yam (*Discorea composite*), "red clover" (*Trifolium pratense*), *Cimicifuga racemosa*, linen, rye and other (Eden, 2001) (Xu et al., 1995).

The major isoflavones are genistein, formononetin, biochanin A and daidzein, which is also metabolized to equol (Bakhit, Klein, Essex-Sorlier, 1994). Due to its characteristics, there is growing interest in the use of phytoestrogens as an alternative to hormone therapy, because it suggests that mainly isoflavones can reduce the effects of menopause and its consequences, with reduced side effects compared to hormonal therapy (Verhoeven et al., 2005). The effects of isoflavones for symptoms of menopause are not well understood, some studies claim that the use of isoflavones reduces the occurrence of hot flashes, while other found no significant results (Cos et al., 2003). Also in hemostasis (Rios et al., 2008) showed a minor side effect of isoflavones compared to classical HRT.

As for osteoporosis studies claim that isoflavones increase bone density, decreasing bone resorption. People using isoflavones have lower rates of PTH (parathyroid hormone) and osteocalcin (Mei; Yeung; Kung, 2001).

However, it is known that soy intake and its metabolites (genistein and daidzein) reduce total cholesterol, mainly by reducing LDL levels, reducing triglyceride levels and increasing HDL (Nestel; Yamashita; Sasahara, 1997) (Washburn et al. 1999; Merritt, 2004).

The phytoestrogens have inhibitory effect on the migration steps and proliferation of smooth muscle cells (Fujio et al., 1993; Shimokado et al., 1994), probably due to the expression of estrogen receptor beta in vascular tissue and other non-reproductive tissues. Moreover, genistein is a specific inhibitor of tyrosine kinase (Akiyama et al., 1987) and isoflavones have an antioxidant role in increasing the resistance of lipoprotein to oxidation. Phytoestrogens extracted from soy *Glicine max* (Andrade et al., 2012) and red clover (Simoncini et al., 2008) reduce *in vitro* the expression of the adhesion molecules VCAM-1, ICAM-1 and E-selectin.

According to these mechanisms of action, phytoestrogens may reduce the development of atherosclerotic plaques (Matti; Adlercreutz; Adlercreutz, 2000).

In addition to the beneficial effects, our group found that isoflavones do not significantly interfere with the blood coagulation, the fibrinolytic system, lipid profile and levels of estradiol and follicle stimulating hormone (RIOS et al., 2008). Our group also found that the isoflavones daidzein and genistein provide protection against cell damage induced by H_2O_2, induce significant increase of NO synthesis and increase the synthesis of prostaglantin E2 in vascular epithelium cells (ECV-304), (Paulo et al., 2009).

Isoflavones

Isoflavones are present in foods derived from soybeans rich in glycosides. Carbohydrate conjugated to the 7-position of isoflavones is subsequently esterified with acetyl or malonyl group at 6 'position. This chemical difference in soy isoflavones may result in different biological activities in different systems. Soybean isoflavone aglycones are absorbed more quickly and in higher doses than their glycoside derivatives in humans (Chien et al., 2006). Besides, the aglycone form has a high biological activity (Lin et al., 2006) implying that products rich in aglycone isoflavones may be more effective than the glycosidic forms in the prevention of chronic diseases such as coronary heart disease or cancer (Lee et al., 2005).

The metabolism of isoflavones varies among individuals and is influenced by other components of the diet, for example, a high amount of carbohydrates causes an increase in the formation of equol. These metabolic compounds produced may have different biological effects and higher estrogenic activity (as equol) than the original isoflavones (Ososki and Kennelly, 2003; Beck et al., 2005).

The bioavailability of isoflavones is an important factor for evaluating the FES supplements. Studies of different preparations resulted in discrepancies between the contents of certain active substances in chemical and biological analysis. The main problem is the low quality of many commercially available products, being necessary to create quality standards for such products (Beck et al., 2005).

In the Faculty of Pharmaceutical Sciences of Ribeirao Preto - USP was produced one biotransformed soybean extract, with aglycone forms of genistein and daidzein.

The extract was obtained by enzymatic action from the fungal enzyme β-glucosidase of *Aspergillus awamori* in soybean flour carefully selected and processed. This treatment generates a methanolic extract with different potential antioxidants which was used therapeutically as an antioxidant in non physiological oxidative stress conditions and used in topical formulations to protect skin against damage caused by absorption of ultraviolet (UV) (Georgetti, 2009). At that time, this biotransformed soybean extract has not been evaluated for their potential estrogenic front of endothelium-derived factors and inflammatory factors.

Endothelial Cell Culture

The use of human endothelial cells in culture systems *"in vitro"* has been indispensable for the standardization of relevant experimental models of inflammation, angiogenesis, wound healing, regulation of blood flow, atherosclerosis, tumor growth and metastasis (Unger et al., 2002). As the basal lamina of the endothelial cells consists of laminin and collagen, collagenase digestion of this blade allows the isolation of endothelial cells to study and culture "*in vitro*" (Mohammad; Bos, 1990). Methods for isolation and culture of primary human endothelial cells from various sources, such as foreskin, bone marrow, placenta, lung and brain have been established (Hewett; Murray, 1993; Schutz; Friedl, 1996), being endothelial cells from human umbilical vein, HUVEC (HUVEC - Human umbilical vein endothelial cell) the more accessible to obtain.

HUVECs have been used extensively to study the biology and pathobiology of human endothelial cell, and most of the knowledge acquired is due to experiments with cultures of this cell. The main advantage of its use is the increased availability of umbilical cord, a relatively simple method of extraction and purity of the population of cells obtained. Endothelial cells exhibit specific cellular characteristics that define the cell type. Primary endothelial cells exhibit a high production of von Willebrand factor (vWF) constituent (Jaffe; Hoyer, Nachman, 1974), adhesion molecules, platelet endothelial cell (PECAM-1, CD31) (Albelda et al., 1990), the two vascular endothelial growth factors, VEGF (vascular endothelial growth factor) and F1T-1 and KDR receptors known to be almost exclusively expressed in vascular endothelial cells (Hewett, Murray, 1996).

In vitro, endothelial cells, typically respond to proinflammatory stimuli such as tumor necrosis factor (TNF), bacterial lipopolysaccharide (LPS) and interleukin-1˙(IL-1), inducing the expression of cell adhesion molecules such as ICAM -1, VCAM-1 and E-selectin (Dustin, Springer, 1998; Bevillacqua et al., 1987; Osborn et al., 1989) at the cell surface.

OBJECTIVES

In this chapter we review and compiled all the results already published and present some unpublished results about our study comparing the effects of an extract of soybean fermented by *Aspergillus awamori* (a β-glucosidases-producing filamentous fungi) on which polyphenol glucosides were bioconverted to aglycone forms with a non-fermented extract on adhesion molecules expression (VCAM-1, ICAM-1 and E-selectin), oxidative stress induced by hydrogen peroxide and on production of nitric oxide, prostaglandin E2 and Endothelin-1 *in vitro* in human endothelial cells (HUVEC).

Materials and Methods

Human Umbilical Cord Collection

Umbilical cords of at least 20 cm long, were collected after normal delivery or cesarean of healthy patients. They were deposited in sterile vials containing saline solution with 1% antibiotic / antimycotic to avoid contamination, not exceeding its use after 6 hours of collection.

Extraction and Cultivation of HUVEC

In laminar flow hood, under sterile conditions, the cord was cleaned externally with 70° alcohol and cannulated with taps of two or three-way at both ends of the umbilical vein, fixed with surgical thread cotton 2.0. The umbilical vein was cleaned inside with sterile PBS to remove any clots present and then, was injected about 20 mL of sterile collagenase type IA (Sigma, St. Louis, MO, US) at 2 mg/mL at 37°C. After 15 minutes, the whole content was collected in a sterile tube containing EBM2 medium (Lonza Group., Switzerland). The cells were centrifuged for 10 minutes at 2500rpm and resuspended in culture medium EBM2.

The percentage of cells not stained by Trypan Blue (Trypan Blue) was considered as a measure of cell viability (Libby et al., 1996) after counting using a Neubauer chamber (VWR Scientifics). HUVEC were grown in 75cm^2 culture flask in EBM2 medium containing 10 ml medium EBM2 with 2% fetal bovine serum, 0.4 IU/ml hydrocortisone, 4 IU/ml rhFGF b (rhFGF - Recombinant Human Fibroblast Growth Factor-B), 1 IU/mL of VEGF (VEGF - Vascular Endothelial Growth Factor Recombinant Human), 1 IU/ml R3-IGF-1 (IGF-1-R 3 - Recombinant long R insulin-like growth factor -1), 1 IU/ml ascorbic acid, 1 IU/mL of rhEGF (rhEGF - Recombinant Human Epidermal Growth Factor), 1 IU / mL gentamicin / amphotericin B and 1 IU/ml heparin. Cells isolated with this technique form confluent monolayers of polygonal cells (cobblestone), analyzed by the expression of von Willebrand factor and / or CD31.

Identification of HUVEC

Endothelial cells exhibit specific cellular characteristics that define the cell type. Primary endothelial cells exhibit a high production of platelet adhesion molecule type 1 (PECAM-1, CD31). Thus, HUVEC were identified by their characteristic polygonal morphology (cobblestone) and the presence of CD31 using PE mouse anti-human CD31 (BD Pharmingen) by flow cytometry (FACS Calibur, BD FACS DIVA software). For this, 1x10^6 cells were placed in flow cytometry tubes, in duplicate, washed with PBS-formaldehyde 1% and incubated with CD31 (10uL), protected from light and under refrigeration (2-8°C) for 15 minutes. They were then washed with 1mL of PBS-formaldehyde 1%, centrifuged at 2500rpm for 10 minutes and resuspended in 500uL of PBS for analysis by flow cytometry. This procedure was repeated three times.

Obtaining Biotransformed Soybean Extracts

The biotransformed soybean extracts were obtained by technique developed at the Faculty of Pharmaceutical Sciences of Ribeirao Preto - USP, in the Quality Control Laboratory of Drug and Cosmetic. Soy beans (*Glycine max*), variety DKO were dried and ground in a hammer mill. The soybean powder was subjected to the extraction process of the oil as standardized by Georgetti, (2009), through a homogenization with isopropanol and centrifugation to degrease the soy flour.

The defatted soy flour was biotransformed by employing an aqueous filtrate containing the fungal spores from the semi-solid culture of *Aspergillus awamori* in culture medium Potato Dextrose Agar (PDA). The biotransformed soybean extracts were obtained from an alcoholic solution - 80% ethanol - (BSEalc.) or an aqueous solution (BSEaq.), filtered and concentrated. Then were performed measurements of genistein (GEN) and daidzein (DAID) by High performance liquid chromatography (HPLC), for measuring the levels of polyphenols and flavonoids.

Analytical Methodology for Quantification of Daidzein and Genistein Using High Performance Liquid Chromatography (HPLC)

The quantification of the products described above was performed by HPLC on a Shimadzu liquid chromatography equipped with a pump, model LC-10AT, and UV-visible detector SPD-10A model. The analyzes were performed on reverse phase Hypersil BDS C18 column (250 x 4.6mm, 5μm particles) Thermo Scientific, coupled to a pre column. The analysis of daidzein and genistein patterns was performed according to the method described by Wang and Murphy (1994), with minor modification. The gradient elution was conducted, and the mobile phase used consisted of 1% acetic acid in acetonitrile (solvent A) and 1% acetic acid in water (solvent B). Following the injection of 20μL of each sample, solvent A was increased from 0% to 35% in 70 minutes. The flow rate used was 1 mL/minute and UV detection was performed at 280nm.

To assess the linearity of the chromatographic method were prepared in triplicate various solutions of daidzein patterns (DAID) and genistein (GEN) (Sigma, St. Louis, MO, US) in methanol 50% in concentrations of 0.1; 0.25; 0.5; 1.0; 5,0μg/mL. The calibration curve was constructed by plotting the concentration on the abscissa axis (x) versus peak height on the ordinate axis (y). Linearity was evaluated by calculating the correlation coefficient and were considered adequate amounts equal to or greater than ≥0,99 (Anvisa, 2003). The identification of the peaks relating to DAID and GEN was performed by comparing the retention times of the sample peaks and those obtained for the standards of DAID and GEN. The amounts of DAID and GEN present in the samples were calculated by the direct relation between the peak heights of the samples with the peak heights obtained from both GEN and DAID standard solutions at a concentration of 0.5mg/mL.

Cell Viability Evaluation

Firstly, it was necessary to set the concentrations of compounds to be used. Were set the concentrations obtained in a study by Simoncini et al. (2005), in which were used the concentrations of 40 ng/mL to genistein and 10ng/mL for daidzein, because these concentrations were found in the plasma of postmenopausal women, receiving standardized "red clover" extract. These concentrations used caused an increase in nitric oxide production by HUVEC (Simoncini et al., 2005).

Furthermore, in previous studies by our group, we use these same concentrations of genistein and daidzein and observe its effectiveness in reducing the expression of adhesion molecules on HUVEC (De Andrade et al., 2012).

Thus, the calculations were performed for both the biotransformed soybean extract (BSE) and soybean extract (SE) had genistein and daidzein in these concentrations. This was established using the concentration of 0.169 mg/mL, according to data provided during the chromatography step, on which was obtained that 0,403mg/mL of the extracts contains 0,2 ug/mL for genistein and 0,24ug/mL for daidzein. Then, the cell viability assays were performed with such concentrations.

The evaluation of the cell viability of HUVEC when exposed to tested compounds was performed by MTT technique (MTT 3- [4,5-dimethylthiazol-2-yl] -2,5-diphenyltetrazolium bromide; thiazolyl blue) and the expression phosphatidylserine (Annexin V) and staining with propidium iodide (PI).

Cell Viability Measurement (MTT 3- [4,5-Dimethylthiazol-2-Yl] -2,5-Diphenyltetrazolium Bromide; Thiazolyl Blue)

To assess the cytotoxic effect of the extracts was used colorimetric method according Mosmann (1983). This method consists in measuring cell viability indirectly by mitochondrial enzyme activity of living cells. In living cells mitochondrial dehydrogenases are able to act on substrates such as 3- (4,5-dimethyl thiazol-2-yl) -2,5-diphenyl tetrazolium bromide (MTT) leading to reductions in the molecule thereby forming the blue formazan. The blue formazan can be solubilized in DMSO and the reading performed by the spectrophotometer at 570nm.

Thus, cells transferred to plates of 96 wells ($2x10^4$ cells/well) were maintained until the confluence of 90% and then treated with E2 (100 pg/mL), BSE and SE (0.169 mg/mL), daidzein (0.1 ug/mL), genistein (0.04 ug/mL) and LPS (1 ug/mL) for 24 hours for cell viability assessment by MTT. The controls used in this test were 1% triton (death control) and cells with no treatment (cell viability control- CV). After 24 hours was added 20uL/well of MTT solution (5 mg/mL).

The plate was then incubated for a further 4 hours in a humidified incubator containing CO2 5% at 37°C. After this period, the culture medium containing MTT was removed and DMSO (200uL/well) was added for 30 minutes at 37°C to dissolve the formazan salts and perform evaluation in a spectrophotometer at 570nm. The tests were performed in triplicate and repeated three times.

Phosphatidylserine Expression (Annexin V) and Staining with Propidium Iodide (PI)

Labeling of Apoptotic Cell by Annexin V

This technique is designed for detection of apoptosis in various cell types, by detection of phosphatidylserine (PS). PS is predominantly observed on the inner surface of the lipid bilayer of the cell membrane, facing the cytosol. In the early apoptotic cells where the cell membrane remains intact but suffers a disorganization, PS is translocated to the outer surface of the bilayer. The appearance of PS on the cell surface is recognized by phagocytes to capture this signal and removes the cell signaled his "suicide" to the environment. Annexin V is a protein that binds to phospholipids and has high affinity for PS in the presence of calcium ions. Changes to this asymmetry of the membrane, which is analyzed by measuring Annexin V to the cell membrane adhesion can be detected before morphological changes associated with apoptosis and early before the loss of membrane integrity. By combining the Annexin V to FITC (fluorescein isothiocyanate) can identify and quantify apoptotic cells by flow cytometry.

Marking of Necrotic Cells by Propidium Iodide

The fluorescent nuclear marker Propidium Iodide (PI) is used to distinguish apoptotic cells from necrotic cells. Propidium iodide is a molecule that intercalates into any DNA provided that the cell membrane is permeable. This property is due to the fact that high molecular weight DNA markers such as PI, are not likely to penetrate the intact cell due to its size and does not mark these apoptotic cells if they not show changes in membrane permeability, that occurs in the final stages of apoptosis. Thus, staining cells simultaneously with FITC-Annexin V (green fluorescence) and PI dye (red fluorescence) allows the discrimination of intact cells, viable (annexin -/PI-) early apoptotic (annexin +/PI-) and late apoptotic or necrotic cells (Annexin +/PI+).

To performing these tests endothelial cells ($5x10^4$ cells/well) were incubated for 24 hours with the treatments, at before mentioned concentrations. They were then trypsinized, washed and incubated with annexin (200μL) binding buffer (10mM HEPES - pH7.4, 150mM NaCl, 5mM KCl, 1mM MgCl2, CaCl2 1.8 mm) and 1.8 ng/ml of Annexin V-FITC (Roche Applied Science) for 15 minutes, protected from light and under refrigeration (2-8°C). At the time of reading, 5μL of propidium iodide were added (100 μg/mL Sigma-Aldrich). The samples were analyzed by flow cytometry (FACScan, Becton Dickinson, San Jose, CA, US) equipped with CellQuest software. The experiment was performed in triplicate and repeated three times.

Evaluation of the Protective Effect against Oxidative Stress

To assess if the compounds tested, in particular soybean extracts (BSE and SE) exert protection against oxidative stress, cells transferred to plates of 96 wells ($2x10^4$ cells/well) were maintained until the confluence of 90% and then treated with different concentrations of H_2O_2 (1mM, 500uM, 250uM, 125uM and 50uM) for 1 hour (Andozia et al., 2010) for definition of hydrogen peroxide concentrations which induce oxidative stress. Subsequently, the HUVEC were pre-treated with 17β-estradiol, BSE, SE, and daidzein with genistein for 24 hours and incubated with H_2O_2 for 1 hour at concentrations of 125uM to 1 mM.

The controls used were cells without any treatment and cells treated with only hydrogen peroxide and the measuring of cytotoxicity was performed by the MTT method as described above.

Adhesion Molecules

Induction of Adhesion Molecules by Lipopolysaccharide (LPS)

Exposure to LPS was tested using 1ug/mL for 12 hours (Andrade et al., 2012). For this, cells transferred to plates of 96 wells ($2x10^4$ cells/well) were maintained until the confluence of 90% and then incubated with LPS 1ug/mL for 12 hours, using untreated cells by LPS in EBM2 medium culture as control.

Evaluation of the Action of BSE in the Expression of Adhesion Molecules (VCAM-1, ICAM-1 and E-Selectin) on the Cell Surface of HUVECs

The expression of adhesion molecules on the cell surface was determined by the EIA method (Enzyme Immuno Assay) (de Catherine et al., 1995; Jonsson, Palmblad, 2001), also called modified ELISA, due to its high sensitivity and ability to detect adhesion molecules directly from the cell surface without the need for lysis of the cells, making use of biotinylated antibodies: VCAM-1 CD106 (R and D Systems), ICAM-1, CD54 (R and D Systems) and E-selectin CD62E (R and D Systems).

Evaluation of the Expression of Adhesion Molecules on the Cell Surface of HUVEC Pretreated with LPS (Therapeutic Action)

In this step, HUVEC were grown in 96-well plates and once confluent, they were exposed to LPS 1ug/mL for 12 hours and treated for 24 hours with BSE (0,169mg/mL), SE (0,169mg/mL), 17β-estradiol (100pg/mL) and genistein with daidzein (40ng/mL and 100 ng/mL, respectively). A control containing only cells in the culture medium was used. After the exposure time of each well, the supernatants were removed and stored for the measurements of the soluble forms of the adhesion molecules and the wells were washed 3 times with PBS, the cells fixed with paraformaldehyde 2% for 30 minutes and blocked with Molico 5% for 1 hour. After blocking were added antibodies conjugated with biotin, anti-VCAM-1, anti-ICAM-1 and anti-E-selectin for 2 hours and streptavidin-peroxidase for 20 minutes, when the reaction was revealed and read at 450 and 540nm (de Catherine et al., 1995; Jonsson; Palmblad, 2001).

Evaluation of the Expression of Adhesion Molecules on the Cell Surface of HUVEC Post-Treated with LPS (Preventive Action)

For this experiment, HUVEC were grown in 96-well plates and once confluent, they were exposed for 24 hours BSE (0,169mg/mL), SE (0,169mg/mL), 17β-estradiol (100pg/mL) and genistein with daidzein (40ng/mL and 100ng/mL respectively), using as control cells in medium without treatment. After the time of exposure to treatment, the cells were stimulated with LPS 1ug/mL for 12 hours. At the end of this period, supernatants were removed and stored for the measurements of the soluble forms of the adhesion molecules and the wells were washed 3 times with PBS and EIA performed on the cell surface, as described above.

Quantification of Soluble Forms of Adhesion Molecules

The quantification of the soluble forms of the adhesion molecules was performed on the stored supernatants by ELISA (Enzyme Linked Immuno Sorbent Assay) using the DuoSet kit ELISA Development System: Human sE-selectin / CD62E (R and D Systems), human ICAM-1 / CD54 (R and D Systems) and Human VCAM-1 / CD106 (R and D Systems). Supernatants were assayed both cultures pretreated with LPS as those of the post-treated cultures with LPS.

Vasoactive Factors

Measure of the Release of Nitric Oxide (NO) by HUVEC

To evaluate the NO production by HUVEC, was initially performed the standardization of DAF-2DA reagent (Sigma, St. Louis, MO, US) by flow cytometry. The probe is incorporated and metabolized by esterases present in HUVEC and NO is proportional to the fluorescent metabolite formed. The analysis of the mean fluorescence intensity was conducted in flow cytometry after 40 minutes from the use of various concentrations of DAF-2DA probe from 10μM to 0.3μM in cell culture ($5x10^5$ cells/tube). Subsequently, HUVEC were incubated with 10μM of 4.5 diaminoflurescein diacetate (DAF-2DA - 4.5 Diaminofluorescein diacetate) for 40 minutes, treated with BSE, SE, daidzein with genistein and 17β-estradiol (E2) in the concentrations described, 30, 60 and 90 minutes and distributed into 96 well plates containing 500 cells/uL in each well, and analyzed by flow cytometry (Guava EasyCyte 8HT - Millipore) for quantification of NO using Incyte-GuavaSoft2 software 2.

The DAF-2DA is a permeable probe into cell membranes. Once inside the cell, diacetate group is hydrolyzed by cytosolic esterases releasing the non-fluorescent compound diaminoflurescein 4.5 (DAF-2 - Diaminofluorescein). The intracellular production of NO takes the DAF-2 fluorescent compound in 4.5 Diaminofluorescein triazole (DAF-2T - Diaminofluorescein triazole) (Leikert et al., 2001). The fluorescence produced in the sample is directly proportional to the intracellular production of NO.

Prostaglandin Dosage (PGE2)

The determination of the production of prostaglandin (PGE2) was performed using the supernatants of cells treated for 24 hours with BSE, SE, 17β-estradiol (E2) and genistein with daidzein, by ELISA using the Parameter Assay Kit Prostaglandin E2 (R and D Systems).

Endothelin-1 Dosage (ET-1)

Determination of production of endothelin-1 (ET-1) was performed using the supernatants of cells treated for 24 hours with BSE, SE, 17β-estradiol (E2) and genistein with daidzein by ELISA using the Quantikine kit Endothelin-1 ELISA (R and D Systems).

Statistical Analysis

The statistical analysis of the results was performed by Graph Pad Prism software (version 4 Prism Graph Pad, US) through one-way analysis of variance (ANOVA) between the groups, with the Tukey's complementary test, considering a significance level of $p<0.05$. Data were expressed as mean ± S.D. All experiments were performed in triplicate and repeated three times.

RESULTS

Collection of the Umbilical Cords, Extraction, Cultivation and Identification of HUVEC

The collection, extraction, cultivation and identification of HUVECs was performed according to the methodology described above and the extracted cells were used until the third passage.

Identification of HUVEC

Figure 1A shows desirable convergence of the human umbilical cord cell cultures obtained by the procedure described in the section Materials and Methods, as well as the morphological identification of the confluent HUVEC monolayer of polygonal cells (cobblestone). Figure 1B shows the analysis of CD31, confirming the identification of cells extracted as HUVEC, demonstrating that 99.7% of the cells showed the presence of CD31 (mean 99.4%).

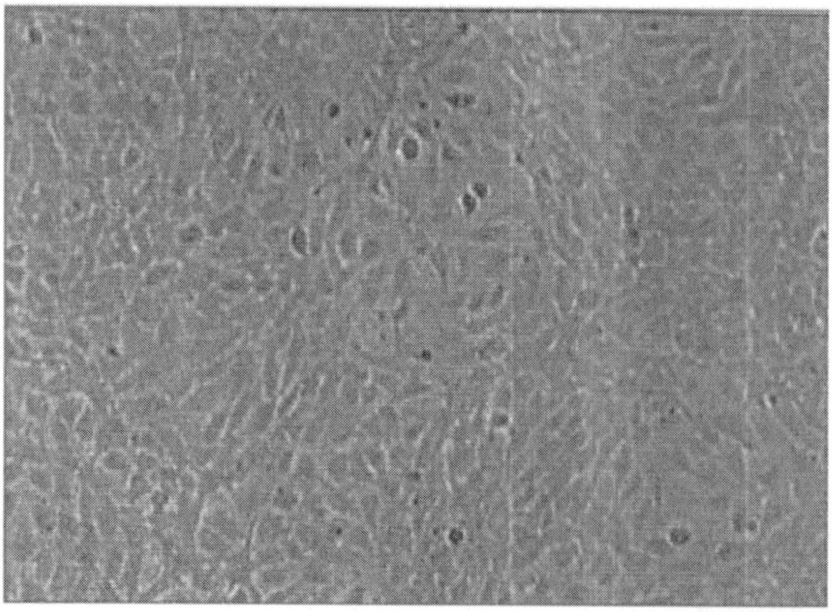

A

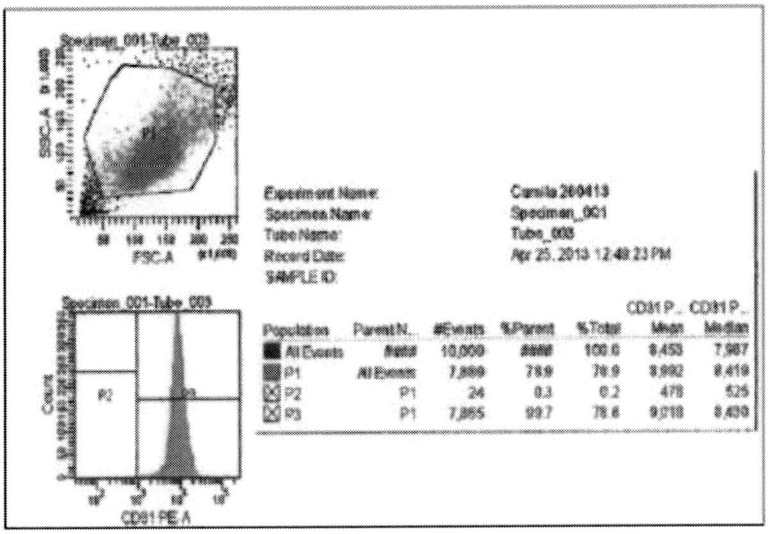

B

Source: De Andrade et al., 2014A, De Andrade et al., 2014B.

Figure 1. A: Culture of HUVEC cells (cobblestone). B: CD31 analysis by flow cytometry.

Chemical Characterization of the Extracts by HPLC

The components present in biotransformed soybean extract (alcoholic and aqueous extract), the crude extract did not metabolized, and the patterns daidzein and genistein were subjected to chromatographic analysis by HPLC, as described and are shown in Figure 2, which follows.

Evaluation of the Effect of BSE on Cell Viability

Measurement of Cell Viability by MTT

The bar graph represents the cell viability of HUVEC cultures (%) after 24 hours.

Cells ($2x10^4$) were cultured in 96-well plates with EBM2 medium and incubated for 24 hours with BSE and SE at concentrations of 0,169mg/mL; daidzein with genistein at the concentrations of 0.1 and 0,04ug/mL, respectively; 17β estradiol in the effective plasma concentration corresponding to the concentration found in the premenopausal period of 100 pg/mL; control of cell viability in cells containing culture medium EBM2 unstimulated control and cell death containing cells incubated with triton 1% in culture medium.

Results were expressed as mean (± standard deviation) of experiments performed in triplicate versus control (CV) analyzed by analysis of variance with $p < 0.05$, and repeated three times. For MTT measurements, the cells showed no decrease in cell viability when exposed to different treatments (Figure 3) and therefore not significant results when treated groups were compared to the control of viability (CV) (Table 1). The significance levels in

Tukey's post-test were p> 0.05 for all treatments compared to CV, demonstrating that the concentrations used did not affect the cell viability.

Phosphatidylserine Expression (Annexin V) and Staining with Propidium Iodide (PI)

The analysis of phosphatidylserine expression and staining with propidium iodide (Figure 4) shows no significant result for the percentage of viable cells (annexin -/PI-), the percentage of cells in early apoptosis (annexin +/PI-) and the percentage of necrotic cells (annexin +/PI+) (table 2), when exposed to treatments.

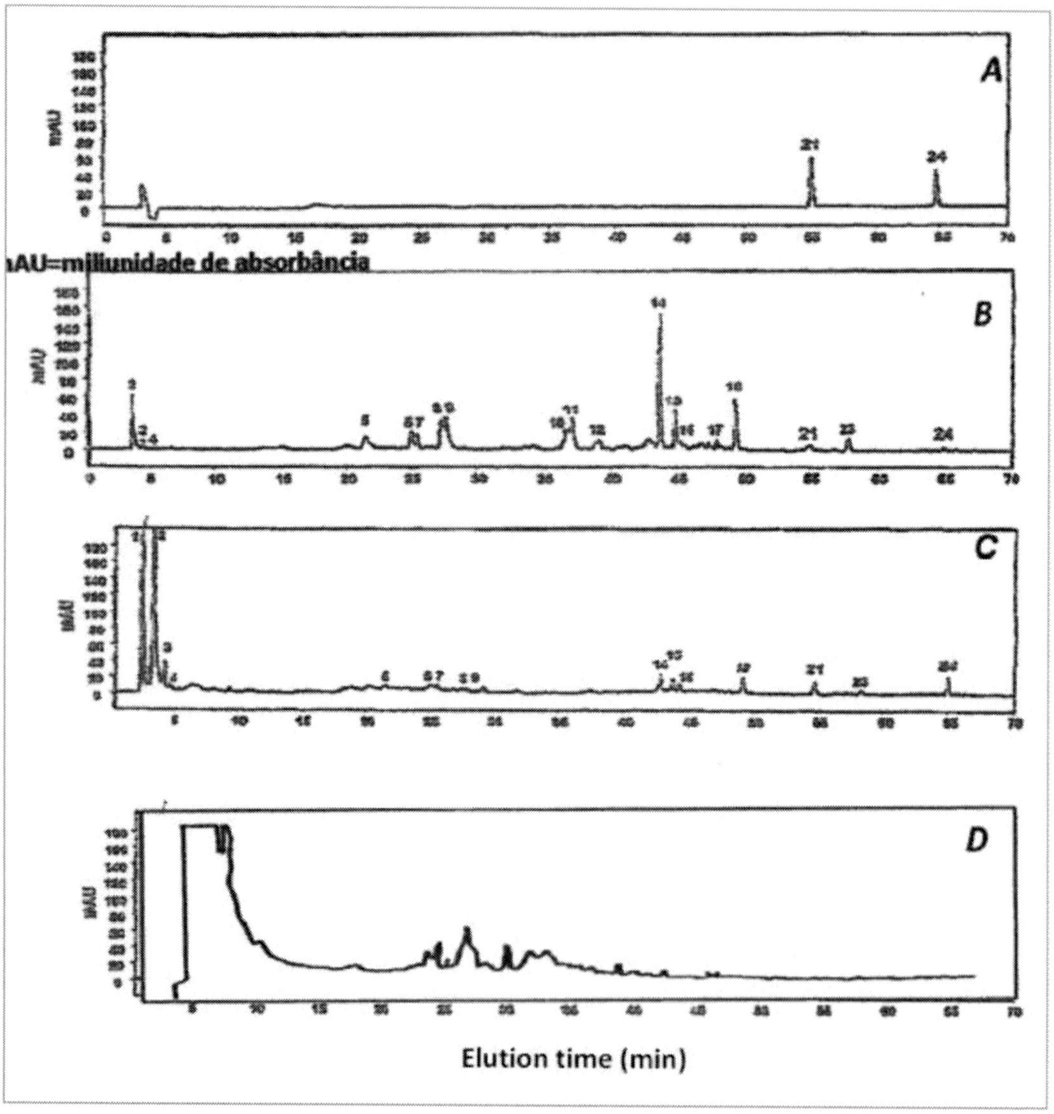

Figure 2. Profile of chromatographic standards, daidzein and genistein, and soy extracts, by HPLC. A - standard chromatogram of daidzein (peak21) and genistein (peak24) at a concentration of 0.5mg/mL; B - SE (alcoholic fraction); C - BSE, alcoholic fraction; D - BSE, aqueous fraction.

Figure 2A corresponds to the chromatogram of the elution patterns of daidzein and genistein represented by the peaks 21 and 24, eluted between 55 and 65 minutes, respectively. Figure 2B corresponds to the SE (alcoholic fraction) elution chromatogram at 80%, which shows that the listed peaks 5, 6, 7, 8, 9, 10, 11, 14, 15 and 18, eluted between 20 and 60 minutes correspond to the phenolic components other than SE isoflavones. The other peaks correspond to glycosylated isoflavones may be: daidzein (peak 12), genistein (peak 16), acetildaidzina (peak 17) and acetilgenistina (peak 23).

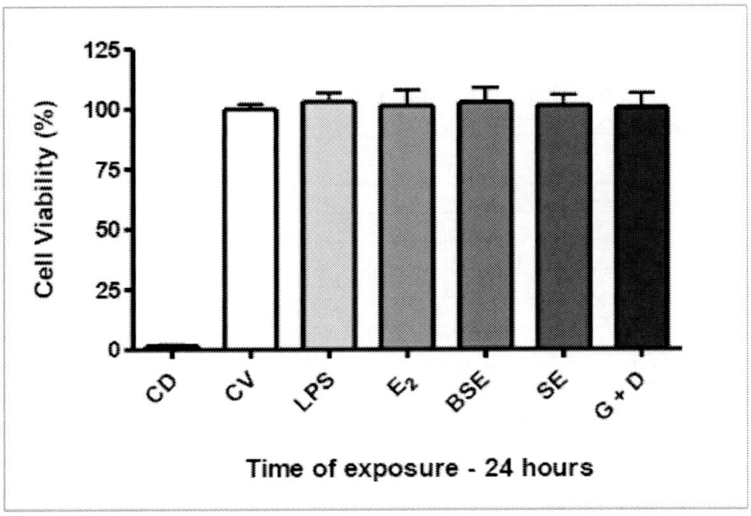

CD-control of death.
CV-control without treatment.
Source: De Andrade et al., 2014A.

Figure 3. Cell viability in HUVEC (%).

Table 1. Results of cell viability measures (%) shown in Figure 3, obtained by MTT expressed as mean ± SD and *p* values for HUVECs exposed to the tested compounds, at concentrations previously described for 24 hours

	Cell Viability – MTT (%)		
	Mean	SD	*p*
CD	1,648	±0,72	-
CV	100,0	±6,12	<0,001*
LPS	103,2	±9,84	NS**
E2	101,4	±29,22	NS**
BSE	102,8	±15,79	NS**
SE	101,4	±12,37	NS**
GD	100,6	±16,16	NS**

*Value obtained by comparing CV versus CD.
**Values obtained by comparing the groups treated versus CV.

Groups were compared with the control group, demonstrating that at concentrations used the substances did not affect the cell viability.

The aglycone isoflavones, daidzein (peak 21) and genistein (peak 24) were identified by comparison with the respective standard retention times (Figure 2A). Figure 2C corresponds to the BSE elution chromatogram, alcoholic fraction, where notes an almost complete reduction of the peaks of phenolic compounds and isoflavone glycosides and also notes a decrease in peak heights of the aglycone isoflavones, daidzein (peak 21) and genistein (peak 24). Figure 2D corresponds to the BSE elution chromatogram, the aqueous fraction. The sums of the peaks from Figures 2B and 2C correspond to the SE and BSE (alcoholic fraction), respectively. The sum of 2D Figure peaks correspond to the BSE (aqueous fraction).

Table 2. Results of Phosphatidylserine Expression (Annexin V) and staining with propidium iodide (PI) in %, shown in Figure 4, obtained by flow cytometry expressed as mean ± SD and *p* values for the compounds tested

	Anexin -/PI-			Anexin +/ PI-			Anexin +/PI+		
	Mean	SD	*p**	Mean	SD	*p**	Mean	SD	*p**
Control	46,75	±3,016	NS	7,043	±1,150	NS	10,61	±1,897	NS
LPS	48,41	±3,525	NS	8,890	±1,499	NS	13,14	±2,294	NS
E2	46,98	±0,69	NS	7,993	±1,060	NS	11,96	±0,575	NS
BSE	52,27	±1,355	NS	6,973	±1,295	NS	9,087	±1,242	NS
SE	51,71	±3,088	NS	10,44	±1,328	NS	11,18	±0,859	NS
GD	50,27	±2,620	NS	9,530	±3,789	NS	11,06	±3,183	NS

**p* values obtained with the comparison of the control group versus treated groups.

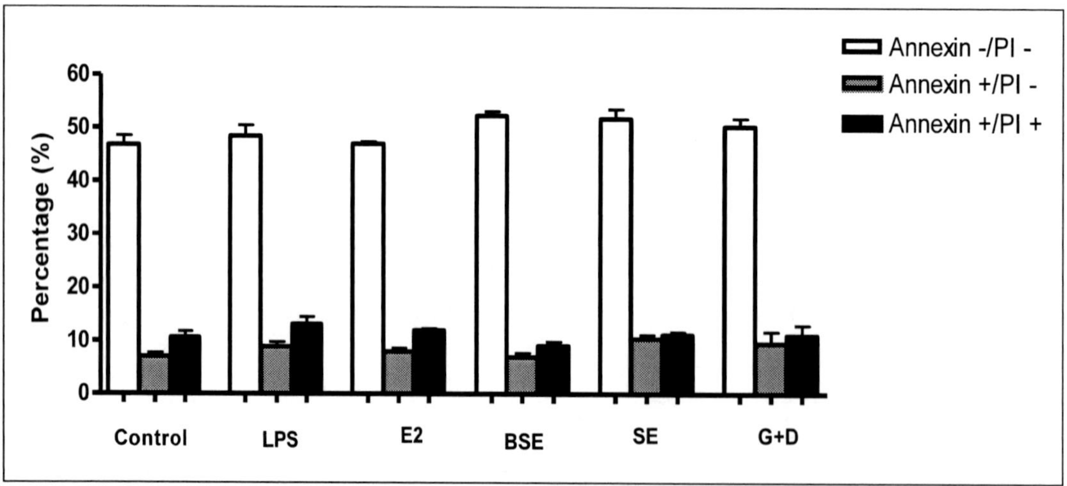

Source: De Andrade et al., 2014B.

Figure 4. The bar graph represents the expression of phosphatidylserine (Annexin V) and staining with propidium iodide (PI) in %. Cells ($5x10^4$) were cultured in 24-well plates with EBM2 medium and incubated for 24 hours with BSE and SE at a concentration of 0,169mg/mL; daidzein with genistein at the concentrations of 0.1 and 0,04ug/mL, respectively; 17β estradiol (E2) at 100pg/mL and control with no treatment. Results were expressed as mean (± standard deviation) of experiments performed in triplicate versus control (C), analyzed by analysis of variance with $p < 0.05$, and repeated three times. The results were analyzed as percentage of cells with no labeling (annexin - / PI -), % of cells stained only with annexin (Annexin + / PI -) and percentage of cells stained with annexin and PI (annexin +/PI +).

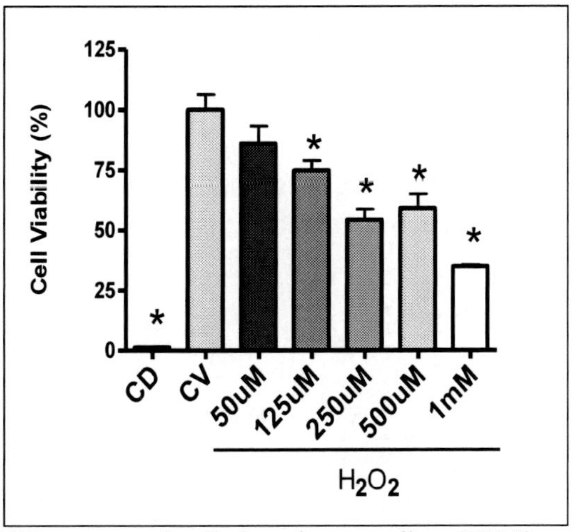

For the control of cell viability was used cells in culture medium EBM2 unstimulated (CV). As death control was used to triton 1% (CD). Results were expressed as mean (± standard deviation) of experiments performed in triplicate versus control (CV) analyzed by analysis of variance with $p < 0.05$, and repeated three times.

Figure 5. The bar graph represents the curve of cell viability (%). HUVEC ($2x10^4$ cells/well) were cultured in 96-well plates with EBM2 medium and incubated for 1 hour with concentrations of H_2O_2 from 50uM to 1mM.

Evaluation of the Protective Effect against Oxidative Stress

Initially, the cells were exposed to various concentrations of H_2O_2 that would induce the oxidative stress. Among the tested hydrogen peroxide concentration (50uM, 125uM, 250uM, 500uM and 1mM) (Figure 5), only the concentration of 50uM presented no significant reduction in cell viability (Table 3).

Table 3. Results of cell viability measures (%) shown in Figure 5, evaluating the oxidative stress induced by hydrogen peroxide at various concentrations, by MTT

	H_2O_2		
	Mean %	SD	*p*
CD	1,290	±0,314	-
CV	100,0	±10,83	<0,001*
50uM	85,96	±7,208	NS**
125uM	74,80	±4,115	<0,001**
250uM	54,22	±4,441	<0,001**
500uM	58,92	±5,923	<0,001**
1mM	34,97	±0,492	<0,0001**

Results expressed as mean ± SD and *p* values. *Value obtained by comparing the CV versus CD. **Values obtained by comparison of groups exposed to hydrogen peroxide versus the CV.

Thus, the four concentrations that showed significant reduction in cell viability were tested in the next stage. According to our results, when the cells were exposed to H_2O_2 125uM, 250 e 500uM (Figure 6) prior exposure of cells to various treatments provided protection against oxidative stress, with significant results for all compounds tested (table 4).

However, the BSE was not more effective than the other treatments, in any of the concentrations tested (p> 0.05).

When the H_2O_2 concentration 1mM was used (Figure 6), prior exposure of cells to compounds did not protect against oxidative stress. This is probably due to the great reduction in cell viability after exposure to 1mM H_2O_2 (Table 4).

Table 4. Means ± SD and *p* values regarding the oxidative stress induced by H_2O_2, demonstrated in figure 6

	H_2O_2 125uM			H_2O_2 250uM			H_2O_2 500uM			H_2O_2 1mM		
	Mean	SD	p^*	Mean	SD	p^*	Mean	SD	p^*	Mean %	SD	p^*
Control	100.0	±11.66		102.0	±6.99		100.0	±10.83		100.0	±2.41	
H_2O_2	71.75	±2.61	<0.05*	59.49	±6.59	<0.001*	60.25	±12.15	<0.01*	34.97	±0.852	<0.001*
$E_2 + H_2O_2$	101.4	±4.29	<0.001**	99.63	±5.30	<0.001**	94.65	±4.68	<0.01**	33.84	±1.706	NS**
BSE + H_2O_2	100.45	±8.61	<0.05**	89.50	±0.48	<0.001**	85.74	±2.55	<0.05**	36.90	±4.152	NS**
SE + H_2O_2	100.7	±12.38	<0.05**	90.84	±2.22	<0.001**	92.05	±3.56	<0.05**	27.88	±3.682	NS**
GD + H_2O_2	95.74	±5.63	<0.05**	91.97	±4.00	<0.001**	93.20	±14.37	<0.01**	26.85	±1.403	NS**

Source: De Andrade et al., 2014A.

*Control x H_2O_2, **Compounds tested + H_2O_2, x H_2O_2.

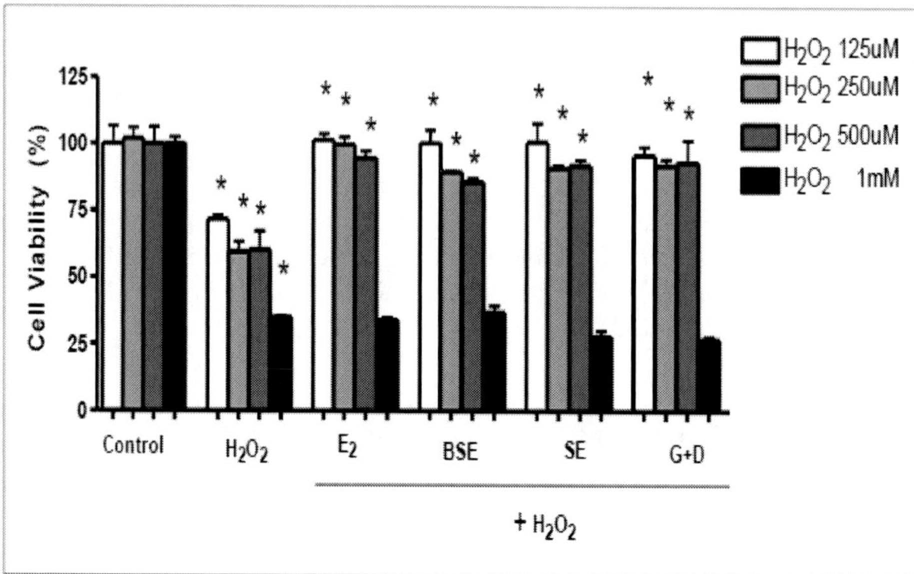

Results are expressed as mean (± SD) of experiments done in triplicates versus controls (H_2O_2) analyzed by analysis of variance with Tukey post test and $p < 0.05$ for comparison of compounds from each other. The experiments were repeated three times (Source:De Andrade et al., 2014A).

Figure 6. Assessment of the protection to oxidative stress. Control - cells without treatment. H_2O_2 - cells exposed only to H_2O_2. The bar graph shows the protection against oxidative stress, in cell viability percentage. HUVEC cultures exposed to H_2O_2 for 1 hour at various concentrations. HUVEC ($2x10^4$ cells/well) were cultured in 96-well plates with EBM2 medium and incubated with 17β-estradiol (E2) (100pg/mL); ESBF (0,169mg/mL); ES (0,169mg/mL) and genistein with daidzein (40ng/mL and 100ng/mL, respectively) for 24 hours and then treated for 1 hour with H_2O_2 at concentrations of 1 mM, 500uM, 250uM and 125uM. Untreated cells exposed only to hydrogen peroxide was used as reducing cell viability control. For the control of cell viability was used cells in culture medium EBM2 (CV).

Table 5. Results of measurements of the adhesion molecules (pg/mL), shown in Figure 7, expressed as mean ± SD and *p* values for HUVECs exposed to LPS 1ug/mL for 12 hours

	VCAM-1 pg/mL			ICAM-1 pg/mL			E-selectin pg/mL		
	Mean	SD	*p**	Mean	SD	*p**	Mean	SD	*p**
Control	2593	±213,1	-	3208	±128,7	-	970,3	±142,8	-
LPS	3119	±146,2	<0,05	4134	±274,7	<0,001	1633	±166,5	<0,01

*Value obtained by comparing the control group versus the group treated with LPS.

Table 6. Means ± SD and *p* values of the dosages on the cell surface and supernatants (pg/mL) of VCAM-1 and sVCAM-1, figure 8A (table 6), ICAM-1 and sICAM-1, figure 8B (table 7) and E-selectin and sE-selectin, figure 8C (table 8)

	VCAM-1 pg/mL			sVCAM-1 pg/mL		
	Mean	SD	*p*	Mean	SD	*p*
Control	1268,0	±154,8	-	1251	±118,6	-
LPS	1935,0	±180,0	<0,001*	1729	±17,13	<0,05*
LPS + E_2	1317,0	±164,2	<0,01**	1423	±15,0	<0,05**
LPS + BSE	865,7	±19,73	<0,001**	1416	±7,45	<0,05**
LPS + SE	1014,0	±42,71	<0,001**	1620	±24,30	<0,05**
LPS + GD	1345,0	±189,5	<0,01**	1378	±99,12	<0,05**

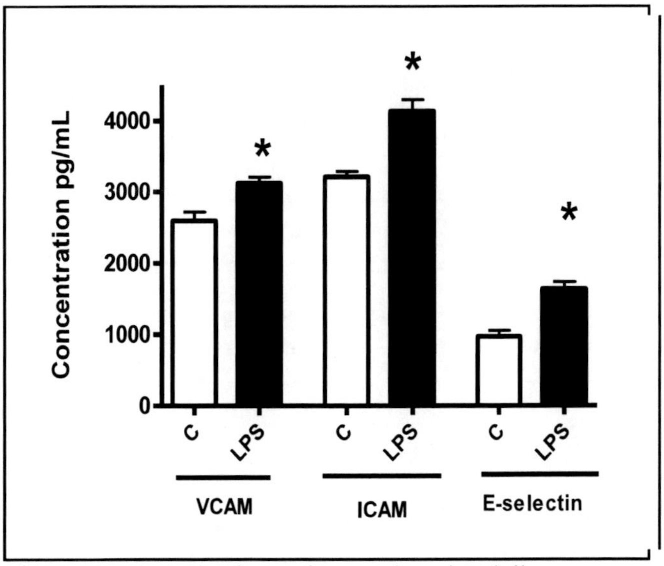

Results are expressed as mean (± SD) of experiments done in triplicates versus control (C), with p <0.05. The experiments were repeated three times.

Figure 7. The bar graph is the standardization of the exposure time and the optimal concentration of LPS as an inflammatory agent in the expression of adhesion molecules by HUVEC. Cells ($2x10^4$ cells/ well) were cultured in 96-well plates with EBM2 medium and incubated with LPS 1ug/mL for 12 hours. Cells not treated with LPS containing only the culture medium were used as control (C) of production of adhesion molecules.

Adhesion Molecules

Induction of Adhesion Molecules by Lipopolysaccharide (LPS)

The exposure of HUVEC to LPS 1ug/mL for 12 hours (Figure 7) significantly stimulated the production of adhesion molecules, VCAM-1, ICAM-1 and E-selectin (Table 5) and therefore the concentration and exposure time used in other phases of the study.

However, BSE, was not more efficient than other phytoestrogens in the reduction of adhesion molecules on the cell surface. Otherwise, BSE was more effective than other phytoestrogens in the reduction of sICAM-1 ($p < 0.05$).

Evaluation of the Expression of Adhesion Molecules on the Cell Surface and in Supernatants of Cultures of HUVEC Pretreated with LPS (Therapeutic Action)

The results showed that at cells where atherogenic process have already started through the expression of adhesion molecules (Figure 8), all treatments tested were able to reduce the expression of VCAM-1 and sVCAM-1 (Table 6), ICAM -1 and sICAM-1 (table 7) and E-selectin and sE-selectin (table 8), demonstrating *in vitro*, therapeutic action.

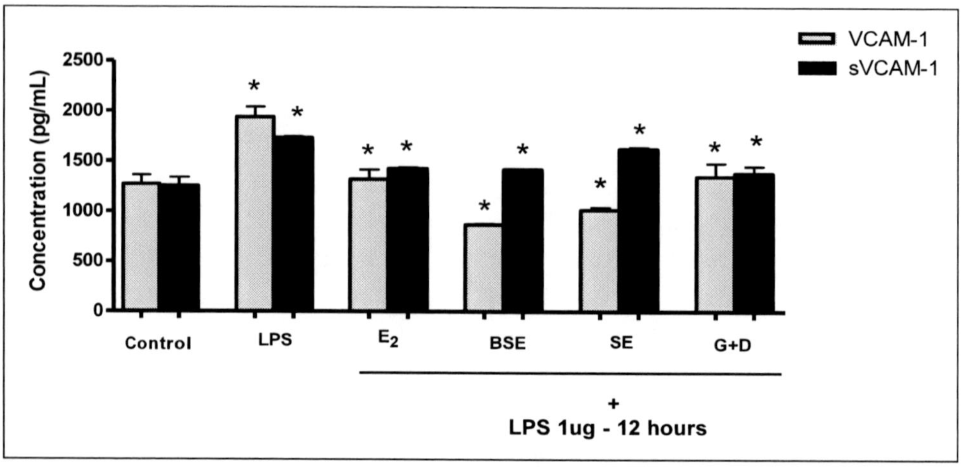

a

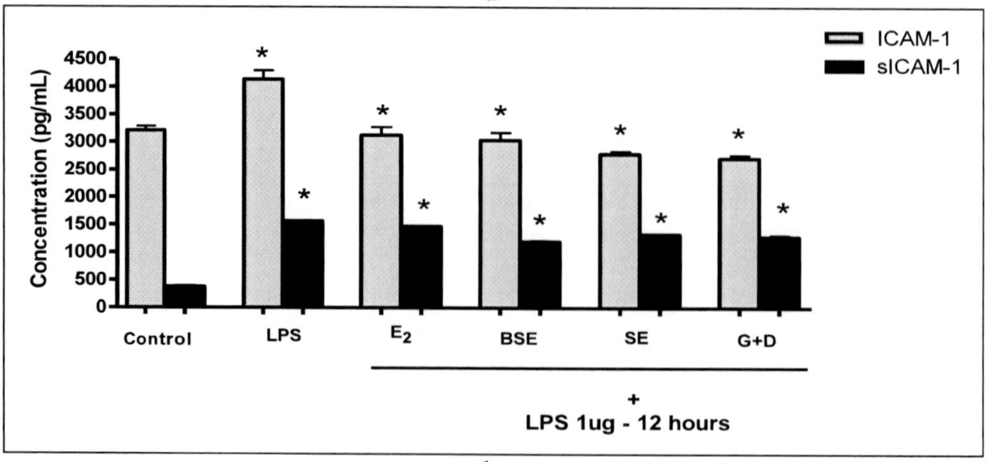

b

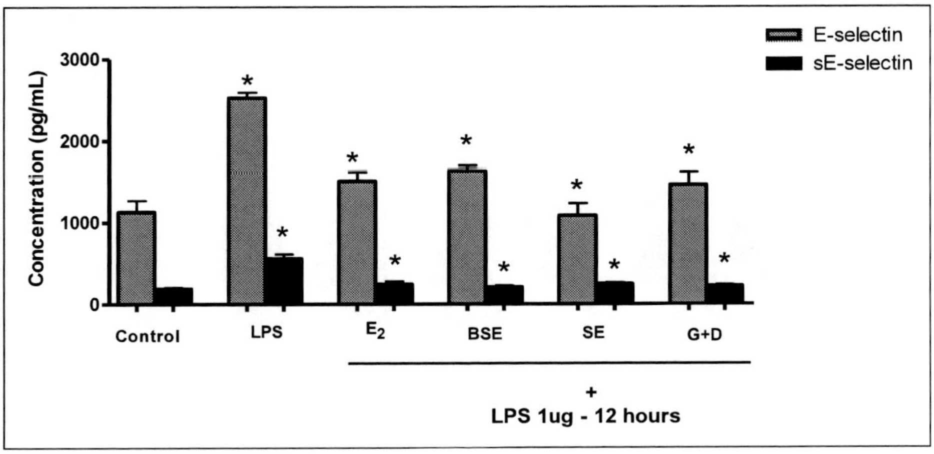

c

Figure 8. Expression of CAM on the cell surface and supernatants (pg/mL) of HUVEC pretreated with LPS (1ug/mL – 12 hours) (therapeutic action) and then treated for 24 hours with BSE (0,169mg/mL), SE (0,169mg/mL), 17β-estradiol (E2) (100pg/mL) with genistein and daidzein (40ng/mL and 100ng/mL, respectively). Control – cells without treatment. LPS – cells exposed only to LPS. Figure 8A: VCAM-1 and sVCAM-1; 8B: ICAM-1 and sICAM-1 and 8C: E-selectin and sE-selectin.

Table 7. Means ± SD and *p* values of the dosages on the cell surface and supernatants (pg/mL) of VCAM-1 and sVCAM-1, figure 8A (table 6), ICAM-1 and sICAM-1, figure 8B (table 7) and E-selectin and sE-selectin, figure 8C (table 8)

	ICAM pg/mL			sICAM pg/mL		
	Mean	SD	*p*	Mean	SD	*p*
Control	3208	±128,7	-	377,2	±20,18	-
LPS	4134	±274,1	<0,001*	1567	±7,51	<0,001*
LPS + E_2	3125	±253,6	<0,001**	1468	±4,91	<0,05**
LPS + BSE	3040	±233,4	<0,001**	1199	±18,38	<0,001**
LPS + SE	2790	±77,41	<0,001**	1328	±3,87	<0,001**
LPS + GD	2713	±97,98	<0,001**	1285	±35,86	<0,001**

Table 8. Means ± SD and *p* values of the dosages on the cell surface and supernatants (pg/mL) of VCAM-1 and sVCAM-1, figure 8A (table 6), ICAM-1 and sICAM-1, figure 8B (table 7) and E-selectin and sE-selectin, figure 8C (table 8)

	E-selectin pg/mL			sE-selectin pg/mL		
	Mean	SD	*p*	Mean	SD	*p*
Control	1127	±240,1	-	187,6	±22,68	-
LPS	2526	±118,6	<0,001*	556,0	±95,21	<0,001*
LPS + E_2	1501	±189,2	<0,001**	242,5	±64,69	<0,001**
LPS + BSE	1625	±132,2	<0,01**	209,8	±29,23	<0,01**
LPS + SE	1080	±259,2	<0,001**	246,7	±28,39	<0,001**
LPS + GD	1455	±272,7	<0,001**	221,1	±25,87	<0,001**

Control x LPS. **Compounds tested + LPS x LPS.

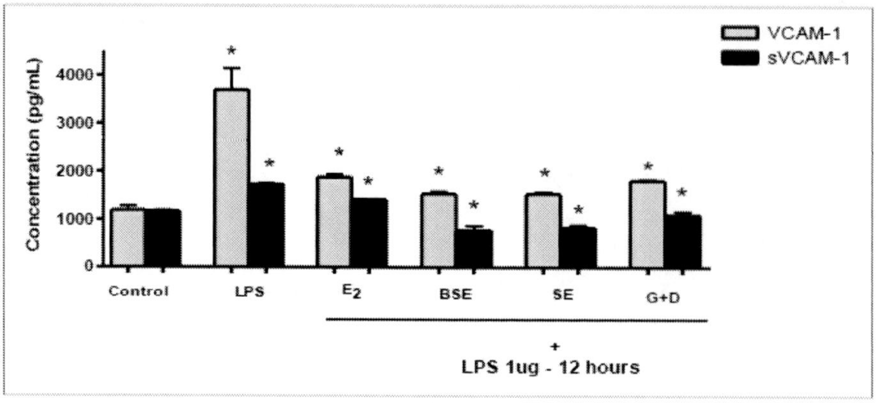

A

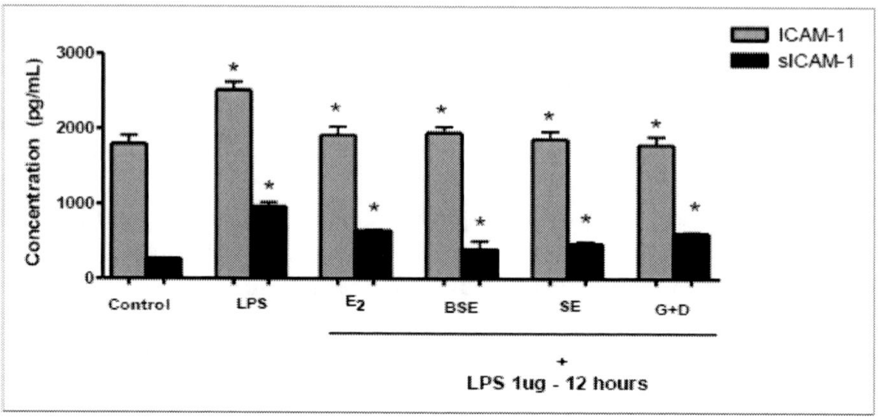

B

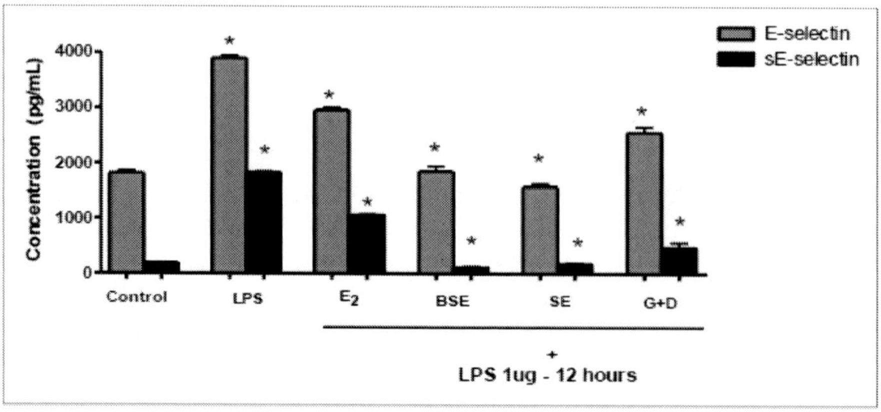

C

Source: De Andrade et al., 2014A.

Figure 9. Expression of CAM on the cell surface and supernatants (pg/mL) of HUVEC treated for 24 hours with BSE (0,169mg/mL), SE (0,169mg/mL), 17β-estradiol (E2) (100pg/mL) and genistein with daidzein (40ng/mL and 100ng/mL, respectively) and then treated with LPS (1ug/mL – 12 hours) (preventive action). Control – cells without treatment. LPS – cells exposed only to LPS. Figure 9A: VCAM-1 and sVCAM-1; 9B: ICAM-1 and sICAM-1 and 9C: E-selectin and sE- selectin.

Tables 9, 10, 11. Means ± SD and *p* values of the dosages on the cell surface and supernatants (pg/mL) of VCAM-1 and sVCAM-1, figure 9A (table 9), ICAM-1 and sICAM-1, figure 9B (table 10) and E-selectin and sE-selectin, figure 9C (table 11), obtained by EIA and ELISA expressed as mean ± SD and P values in HUVECs pretreated with compounds at the concentrations tested for 24 hours and then stimulated with LPS 1ug / mL for 12 hours

Table 9	VCAM-1 pg/mL			sVCAM-1 pg/mL		
	Mean	SD	*p*	Mean	SD	*p*
Control	1190	±143.2	-	1165	±32.32	-
LPS	3690	±776.2	<0.001*	1729	±17.06	<0.01*
E_2 + LPS	1879	±93.70	<0.001**	1406	±9.76	<0.05**
BSE + LPS	1528	±85.46	<0.001**	776.4	±122.2	<0.001**
SE + LPS	1527	±78.42	<0.001**	832.8	±73.68	<0.001**
GD + LPS	1808	±33.50	<0.001**	1099	±94.90	<0.001**

Table 10	ICAM-1 pg/mL			sICAM-1 pg/mL		
	mean	SD	*p*	mean	SD	*p*
Control	1799	±192.8	-	264.8	±0.49	-
LPS	2521	±178.1	<0.01*	965.0	±70.91	<0.05*
E_2 + LPS	1916	±195.8	<0.05**	648.8	±11.36	<0.05**
BSE + LPS	1948	±140.0	<0.05**	398.5	±150.6	<0.01**
SE + LPS	1862	±180.3	<0.01**	469.1	±41.92	<0.01**
GD + LPS	1789	±180.4	<0.01**	611.4	±22.65	<0.05**

Table 11	E-selectin pg/mL			sE-selectin pg/mL		
	mean	SD	*p*	mean	SD	*p*
Control	1814	±83.08	-	187.6	±22.68	-
LPS	3898	±70.47	<0.001*	1837	±25.68	<0.001*
E_2 + LPS	2957	±91.00	<0.001**	1056	±30.37	<0.001**
BSE + LPS	1855	±162.8	<0.001**	113.6	±30.04	<0.001**
SE + LPS	1572	±96.44	<0.001**	186.4	±12.07	<0.001**
GD + LPS	2560	±172.5	<0.001**	473.5	±164.00	<0.001**

Source:De Andrade et al., 2014A.

*Value obtained by comparing the control group versus group treated with LPS.

**Values obtained by comparison of the group treated with LPS only treated groups versus previously tested compounds and LPS.

Evaluation of the Expression of Adhesion Molecules on the Cell Surface and in Supernatants of Cultures of HUVEC Post-Treated with LPS (Preventive Action)

The results showed that in cells where the atherogenic process have not yet started through the expression of adhesion molecules (Figure 9), all treatments carried out prior to the stimulus were able to reduce the expression of VCAM-1 and sVCAM-1 (Table 9), ICAM

-1 and sICAM-1 (table 10) and E-selectin and sE-selectin (table 11), demonstrating in vitro, preventive action.

BSE was not more efficient than other phytoestrogens in the reduction of adhesion molecules on the cell surface. Otherwise, BSE was more effective than other phytoestrogens in the reduction of sICAM-1 ($p < 0.001$).

Vasoactive Factors Endothelium

Measure of the Release of Nitric Oxide (NO) by HUVECs

Standardization of the Concentration of DAF-2DA Probe for Measurement of NO in HUVEC

Initially, it was standardized concentration of DAF-2DA probe used (Figure 10).

Dosages Nitric Oxide (NO)

The results of the nitric oxide measurement by flow cytometry (Figure 11) showed that only groups exposed to BSE for 60 and 90 minutes showed increased nitric oxide production by HUVEC (Table 12).

Prostaglandin Dosage

The results of the dosages of prostaglandin E2 (Figure 12) showed that only the cells exposed to BSE for 24 hours presented an increase in prostaglandin E2 production in HUVEC (Table 13).

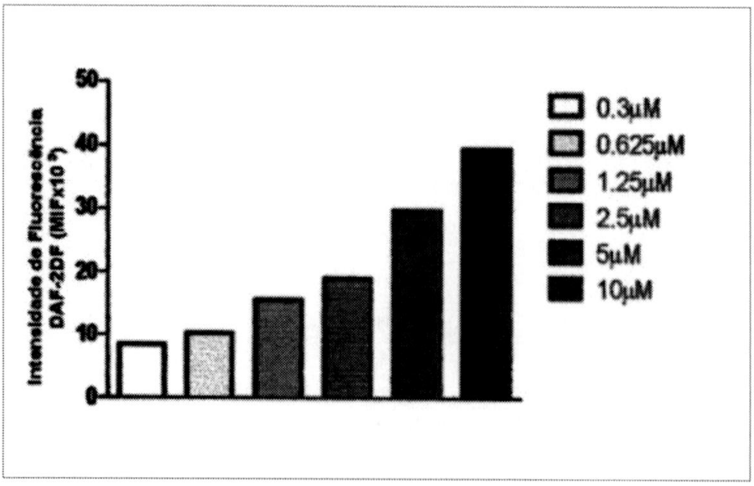

The bar graph represents the mean fluorescence intensity on the development and metabolism of the probe at different concentrations: 10uM to 0,3uM to determine the production of NO by endothelial cells.

Figure 10. Quantification of NO by flow cytometry on HUVEC. HUVEC (5×10^5 cells/well) were incubated for 40 minutes with DAF-2DA probe. After this period the mean fluorescence intensity was evaluated by flow cytometry.

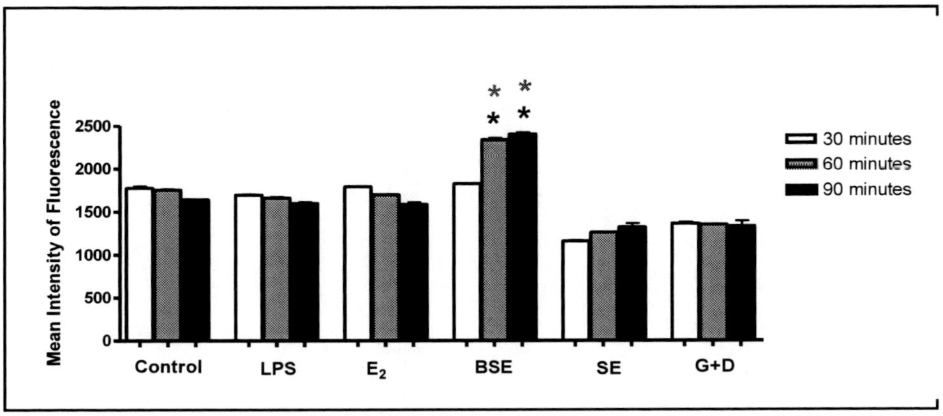

Source: De Andrade et al., 2014B.

Figure 11. Bar graph represents the production of NO by HUVEC, by flow cytometry. (1.75 x 10^5 cells/ well) were incubated with DAF-2DA probe (1uM) for 40 minutes and then treated with BSE (0,169mg/ mL), SE (0,169mg/mL), 17β-estradiol (E2) (100pg/mL) and genistein with daidzein (40ng/mL and 100 ng/mL respectively) for 30, 60 and 90 minutes. After this period the mean fluorescence intensity was evaluated by flow cytometry.

Table 12. Results of measurements of nitric oxide (NO), shown in Figure 11, obtained by flow cytometry expressed as mean ± SD and *p* values in HUVECs treated with the compounds tested at the concentrations described for 30, 60 and 90 minutes

	NO 30'			NO 60'			NO 90'		
	Mean	SD	p^*	Mean	SD	p^*	Mean	SD	p^*
Control	1774	±34,75	-	1752	±17,04	-	1638	±10,34	-
LPS	1692	±14,44	NS	1658	±28,15	NS	1594	±29,15	NS
E_2	1788	±3,57	NS	1695	±12,09	NS	1584	±39,48	NS
BSE	1822	±14,45	NS	2334	±36,76	<0,001	2401	±38,37	<0,001
SE	1157	±13,46	NS	1257	±9,74	NS	1317	±74,90	NS
GD	1359	±31,77	NS	1346	±15,48	NS	1330	±85,17	NS

$^*p$ values obtained by comparing the control group versus treated groups.

Table 13. Results of measurements of prostaglandin E2 (pg/mL) shown in Figure 12, obtained by ELISA expressed as mean ± SD and *p* values in HUVECs treated with the compounds tested at the concentrations described for 24 hours

	Prostaglandin E2 (pg/mL)		
	Mean	SD	p^*
Control	199,8	±76,81	-
LPS	348,9	±40,99	NS
E_2	211,1	±114,2	NS
BSE	1410	±301,6	<0,001
SE	327,0	±56,79	NS
GD	105,9	±78,48	NS

$^*p$ values obtained by comparing the control group versus treated groups.

Endothelin-1 Dosage

Results from dosages of endothelin-1 (Figure 13) showed that all treatment groups showed increases in production of endothelin-1 in HUVEC (Table 14). However, the increase caused by BSE was lower than the increase caused by SE.

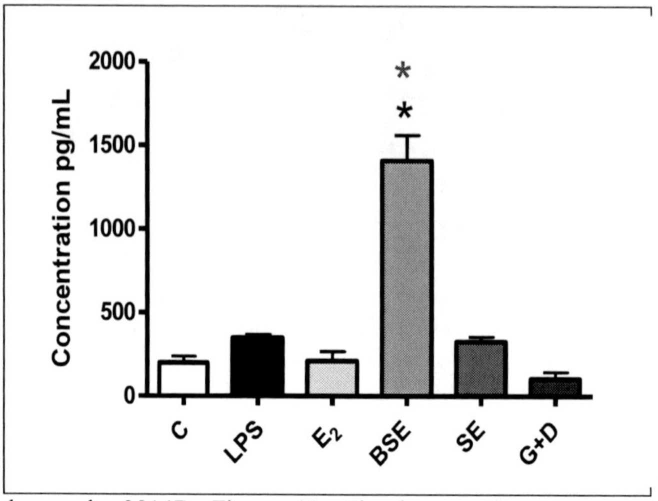

Source: De Andrade et al., 2014B. Figure 12. The bar graph represents the concentration of Prostaglandin E2 (pg/mL) in supernatants from cultures of HUVECs treated with BSE (0,169mg/mL), SE (0,169mg/ mL), 17β-estradiol (E2) (100pg/ mL) and genistein with daidzein (40ng/mL and 100ng/mL respectively) for 24 hours by ELISA using the Parameter Prostaglandin E2 Assay kit (R and D Systems).

Table 14. Results of measurements of prostaglandin E2 (pg/mL) shown in Figure 13, obtained by ELISA expressed as mean ± SD and *p* values in HUVECs treated with the compounds tested at the concentrations described for 24 hours

	Endotelina-1 (pg/mL)		
	Mean	SD	p^*
Control	218,2	±58,49	-
LPS	272,1	±31,92	<0,05
E_2	370,8	±17,55	<0,001
BSE	358,9	±28,63	<0,001
SE	452,3	±28,01	<0,001
GD	334,3	±6,58	<0,001

* *p* values obtained by comparing the control group versus treated groups.

Discussion

Atherosclerosis (AS) is a chronic inflammatory disease and, therefore, the inflammatory process is crucial in all stages of atherosclerosis. Under normal conditions, endothelial cells inhibit the adhesion of leukocytes and platelets to vascular surfaces.

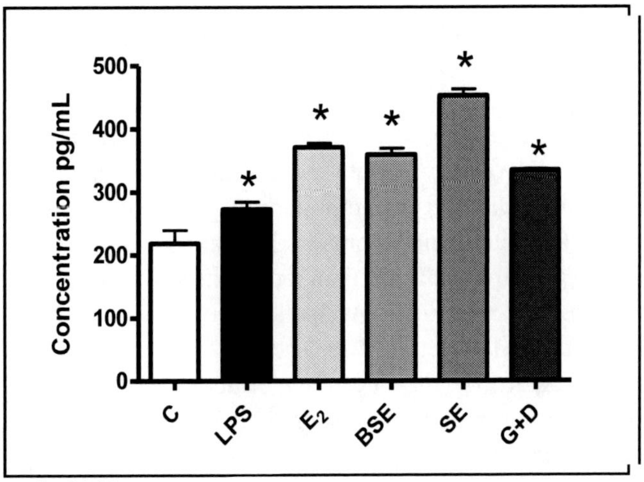

Source: De Andrade et al., 214B.

Figure 13. Bar graph represent the concentration of endothelin-1 (pg/mL) in supernatants from cultures of HUVECs treated with BSE (0,169mg/mL), SE (0,169mg/mL), 17β-estradiol (E2) (100pg/mL) and genistein with daidzein (40ng/mL and 100ng/mL respectively) for 24 hours by ELISA using the Quantikine ELISA kit Endothelin-1 (R and D Systems).

Risk factors such as hyperlipidemia, toxic or infectious agents (Matti; Adlercreutz, Adlercreutz, 2000), cause endothelial injury by stimulating the cells and initiating an inflammatory response in the arterial wall (Lteif; Han and Mather, 2005).

The initial stage is characterized by an increase in the levels of inflammatory cytokines, oxidative stress and expression of adhesion molecules in the vascular endothelium.

Oxidative stress plays an important role in the cell damage and the reactive oxygen species (ROS) are involved in the apoptosis of endothelial cells (EC) (Gong et al., 2010). The apoptosis is a programmed cell death process in which injured cells are eliminated for the maintenance of homeostasis.

Endothelial cells serve as a barrier between the tissues and circulation. Apoptosis of endothelial cells causes the disruption of this barrier and is implicated in numerous pathophysiological processes, such as, angiogenesis, thrombosis, atherosclerosis, hypertension, heart failure and diabetic vasculopathy (Gong et al., 2010). Hydrogen peroxide (H_2O_2) is one of the main endogenous sources of ROS (Nohl et al., 2003), therefore, it is broadly used in the induction of oxidative stress *in vitro* models (Yang et al., 2006).

Several adhesion molecules (CAMs) have been identified in endothelial cells, such as, VCAM-1 (vascular cell adhesion molecule), ICAM-1 (intercellular adhesion molecule) (Rimbach et al., 2008), (Berliner et al., 1995) and E-selectin (de Andrade et al., 2012). The adhesion molecules expression in endothelial cells is stimulated by interleukin (IL)-1β, tumor necrosis factor (TNF)-α and lipopolysaccharide (LPS), which serve as signaling molecules to facilitate transactivation of nuclear factor (NF) κB (Bowie et al., 1997) (May et al., 1996), (Adamsnon et al., 1996). The expression of adhesion molecules is closely related to the regulation of NFκB (Rimbach et al., 2008).

The endothelium also produces nitric oxide (NO), synthesized by endothelial nitric oxide synthase (eNOS), which is a vasoprotetor and anti-atherosclerotic agent. An eNOS isoform is constitutively expressed and can be activated and inhibited by phosphorylation by various protein-kinases (Vera et al., 2007) (Cho et al., 2011).

Hence, any substance that reduces oxidative stress, suppress adhesion molecules and increases the production of nitric oxide may be used as a therapeutic agent against the development of atherosclerosis.

The isoflavones present in soy products may show wide variation in their quantity and quality according to the variety of soy and its method of production (Wang; Murphy, 1994). Aglycone isoflavones (active biological compounds with estrogenic activity), are absorbed faster and in larger quantities than their precursors glycosides (Chien et al., 2006). Moreover, the aglycone forms have greater activity (Lin et al., 2006), implying that aglycone-rich isoflavones may be more effective than glycoside-rich ones in the prevention of chronic diseases such as atherosclerosis (Lee et al., 2005).

How aglycone-rich isoflavones may be more effective than glycoside-rich ones in the prevention and treatment of chronic diseases such as AS, (Lee et al., 2005), the *Glycine max* soybean extract biotransformed by the fungus *Aspergillus awamori* may have a greater estrogenic potential in face of pro-inflammatory factors (De Andrade et al., 2014A).

The effects of this extract also was compared with soybean extracts (SE) without biotransformation, 17β-estradiol (E2) and genistein with daidzein in human endothelial cells (HUVECs) *in vitro*.

Abnormalities in the structure and function of the endothelial cells may contribute to the vascular wall disorders, such as thrombosis, atherosclerosis and vasculitis. Therefore, HUVECs are most commonly used in cell culture models to elucidate mechanisms of cardiovascular disease *in vitro* (Potdara and Kavdia, 2009).

For this, initially evaluated if compounds studied did not interfere in cell viability, by the MTT assay and the if these compounds would induce apoptosis or necrosis, by the expression of phosphatidylserine and staining with propidium iodide.

According to the results, the compounds studied did not presented interferences in cell viability (Figure 3, table 1) (De Andrade et al., 2014(A) and did not induce cell apoptosis or necrosis (Figure 4, table 2) (De Andrade, et al., 2014(B). The next stage was to evaluate the protection against damage caused by oxidative stress induced by hydrogen peroxide (H_2O_2). Initially evaluated the concentrations that would be capable to induce oxidative stress (figure 5, table 3). Then, was evaluated the protective effect of the studied compounds against oxidative stress (figure 6, table 4) (De Andrade et al., 2014A).

The previous exposure of the cells to the several treatments granted protection against the oxidative stress at 125, 250 and 500uM of H_2O_2 (figure 6), presenting significant results for all of the compounds tested (table 4). Nevertheless, the BSE was not found more effective in comparison to the other treatments for any of the concentrations tested ($p>0.05$). When using a concentration of 1mM of H_2O_2 (figure 6), the previous exposure of the cells to the compounds did not protect them against oxidative stress (table 4). Thus, there was protection against oxidative stress when cell viability was higher than 50% and this capability was lost when less than 50% of the cells were viable. The reduction in cell viability occurred in a dose-dependent manner (figure 5 and table 3), which was also verified by Gong et al. (2010). In the assessment of antioxidant effects of the equol in BAECs, Chung et al. (2008) found 39.10 (±3.67%) of cell viability in the control group when exposed to 1mM of H_2O_2, during 1 hour, without increased viability in the group treated with daidzein, which agrees with the results of this study, obtained in HUVECs (34.97 ± 0.492%). Nevertheless, greater cell viability was also found in the group treated with estradiol, which did not occur in this study. Our results agree with those of Andozia et al. (2010), which found 38.5 (±4.9%) of cell viability in ECV304, treated with H_2O_2 at 1mM for 1 hour, and did not find significant results in cells that were pre-treated with E2 at 100pg. Therefore, in HUVECs, the BSE granted protection against the oxidative stress, but was not found more efficient than the other compounds tested (De Andrade et al., 2014 A).

The subsequent step was the evaluation of the expression of CAMs, VCAM-1, ICAM-1 and E-selectin on the cell surface and in the supernatants of cultures of HUVEC.

HUVECs have been used extensively to study the biology and pathobiology of human endothelial cells. In vitro, endothelial cells, typically respond to proinflammatory stimuli such as tumor necrosis factor (TNF), bacterial lipopolysaccharide (LPS) and interleukin-1 (IL-1), inducing the expression of cell adhesion molecules such as ICAM-1, VCAM-1 and E-selectin on the cell surface (Dustin, Springer, 1998; Bevillacqua et al., 1987; Osborn et al., 1989).

Lipopolysaccharide (LPS), a major component of the cell wall of gram-negative bacteria, is a potent activator of immune cells and inflammatory system (Wolk et al., 2003). Once circulating LPS binds to a binding protein LPS PLB (PLB, ligand binding protein) forming a complex which binds with high affinity to a cell surface receptor of macrophages and monocytes, CD14 (Verbon et al., 2001). A transmembrane signal is generated and the result is cellular activation of NF-kβ and secretion of inflammatory cytokines. CD14 soluble can also bind to the complex LPS/PLB and activate endothelial cells, which do not express CD14 on the surface (Noubir; Hmana; Reiner, 2004).

It is known that ICAM-1 is basally expressed at low levels in vascular endothelium, but can be induced to high levels when stimulated with LPS. Studies have shown that stimulation of ECV304 with 1ug of LPS for 1, 2 and 4 hours induces the production of mRNA for ICAM-1 (Antonelli et al., 2001), and measurement by flow cytometry, demonstrated that 8 hours after LPS stimulation, an increase in the expression of ICAM-1 happened (Moreno, 2001).

E-selectin is normally absent in normal conditions and its transcription is induced by cytokines such as interleukin-1, tumor necrosis factor, and LPS (Tedder et al., 1999). There is induction of the expression of E-selectin in ECV 304 when stimulated with 1ug of LPS, reaching maximum values after one hour of incubation, when analyzed by flow cytometry (Moreno, 2001). Studies have found increased expression of VCAM-1 and ICAM-1 in established atherosclerotic lesions and development (Zakynthinos, Pappa, 2009), whereas genetic deficiencies in VCAM-1 and ICAM-1 are associated with decreased atherosclerosis (Nageh et al. 1997) transforming VCAM-1 and ICAM-1 in markers of early events of atherosclerosis. It is known that phytoestrogens exert a number of beneficial effects on cardiovascular risk markers an inverse relationship occurring between foods containing soy-based diets and the risk of atherosclerosis (Wenzel et al., 2008). Phytoestrogens are capable of reducing the soluble forms of VCAM-1, ICAM-1 and E-selectin in the peripheral blood of postmenopausal women (COLACURCI, 2005) and in cultured HUVECs (De Andrade et al., 2012), as well as the reducing the expression of VCAM-1, ICAM-1 (Simoncini et al., 2008) and VCAM-1, ICAM-1 and E-selectin on the cell surface of HUVECs stimulated with LPS (De Andrade et al., 2012), reinforcing the hypothesis that these compounds may exert anti-inflammatory actions in the endothelial cells.

Hence isoflavones can act in the prevention of chronic diseases such as atherosclerosis and how any substance that reduces oxidative stress, suppress adhesion molecules and increases the production of nitric oxide may be used as a therapeutic agent against the development of atherosclerosis, is possible that isoflavones can act as therapeutic agent and/ or preventive agent. Previous studies (Andrade, 2009) demonstrated that phytoestrogens can act as preventive and therapeutic agent, reducing the expression of VCAM-1, ICAM-1 and E-selectin and its soluble forms in HUVEC and ECV-304, stimulated with LPS. Otherwise, *in vivo,* it is more probably that endothelial cells are already activated, which occurs in men and postmenopausal women.

Thus, the expression of adhesion molecules was evaluated testing the therapeutic and the preventive action of the BSE. We tested if in previously activated HUVEC, BSE would be able to reduce the adhesion molecules, which would characterized it therapeutic action, allowing them to be used in a possible treatment of atherosclerosis. We also tested if in HUVEC not activated, the BSE can act as preventive agent.

Firstly, the expression of adhesion molecules by LPS was confirmed using it at concentration of 1ug/mL for 12 hours (Figure 7, table 5) (Andrade, 2009).

The results demonstrated that in cells where the atherogenic process has not yet started through the expression of CAM (figure 9, tables 9, 10 and 11), all treatments performed prior to the beginning of the stimulus were capable of reducing the expression of VCAM-1 and sVCAM-1 (figure 9A, table 9), ICAM-1 and sICAM-1 (figure 9B, table 10) and E-selectin and sE-selectin (figure 9C, table 11), thus demonstrating preventive action *in vitro.*

The BSE presented greater efficacy than the other compounds in the reduction of sICAM-1 ($p<0.01$) (De Andrade et al., 2014A). When BSE was tested in HUVEC where the atherogenic process already has started through the expression of CAMs (figure 8, tables 6,7,8), the results demonstrated that all treatments performed after to the beginning of the stimulus were capable of reducing the expression of VCAM-1 and sVCAM-1 (figure 8A, table 6), ICAM-1 and sICAM-1 (figure 8B, table 7) and E-selectin and sE-selectin (figure 8C, table 8), thus demonstrating therapeutic action *in vitro.* The BSE presented greater efficacy than the other compounds in the reduction of sICAM-1 ($p<0.001$).

The results showed also match the expression of adhesion molecules on the cell surface and their release into the supernatant, being the tests of soluble forms of CAMs are used for studies "*in vivo*", where the evaluation of the expression of these molecules in cell surface, presents greater difficulties. In addition, the results obtained agree with studies that demonstrated that genistein acts as an anti-inflammatory and antiatherogenic agent in human endothelial cells, reducing the expression of VCAM-1, ICAM-1 (De Andrade et al., 2012) (Simoncini et al., 2008) and E-selectin (De Andrade et al., 2012).

The last step consisted in the assessment of endothelium-derived factors. Dosages by nitric oxide, prostaglandin E2 and endothelin-1 were performed.

As a result, we found that all compounds tested increased the production of ET-1 (Figure 13, table 14). However, the increase caused by BSE was significantly lower ($p <0.001$) than the increase observed with ES (De Andrade et al., 2014B). Furthermore, after standardization of DAF-2DA probe concentration- Figure 10) we found that only BSE caused increased production of nitric oxide (Figure 11 and Table 12) and increased production of prostaglandin E2 (figure 12 and table 13) (De Andrade et al., 2014B).

Several cell types can produce and endothelin-1 release to the extracellular medium, such as cardiac myocytes, smooth muscle cells, fibroblasts and epithelial cells. However, the main biological source of endothelin-1 production is the endothelium (Yanagisawa et al., 1988). Endothelin-1 (ET-1) binds to two types of receptors: ETA and ETB.

It is traditionally accepted that the ETA receptor, is expressed in smooth muscle cells, mediates the proliferative and vasconstritive response in smooth muscle cells, associated with hypertension, diabetes and other cardiovascular diseases, while the ETB receptor, is expressed in smooth muscle cells and cells endothelial, is considered responsible for promoting endothelial synthesis of vasodilators nitric oxide and prostacyclin and the clearance of circulating endothelin-1 (Thorin and Webb, 2010).

80% of endothelin-1 produced by endothelial cells diffuses through the basal side of the vessel wall and binds to the ETA and ETB receptors of smooth muscle cells, leading to vasoconstriction (Wagner et al., 1992). The ET-1 that diffuses into the plasma activates ETB receptors on endothelial cells.

The endothelial cells express only ETB receptors, which perform two main actions: release of nitric oxide (NO) and prostaglandins and ET-1 clearance from the circulation. (Bagnal et al., 2006; Kelland et al., 2010a; Kelland et al., 2010b). Thus, the ETB activation stimulates the release of prostacyclin, cyclooxygenases (Tirapelli et al., 2005), prevents apoptosis (Quehenberger et al., 2000) and increases the production of nitric oxide by endothelial nitric oxide synthase (eNOS), which produces vasodilation (Tsukahara et al., 1994). In rats, strong expression of ET-1 increases oxidative stress, promotes local inflammation and induces vascular hypertrophy (Amiri et al., 2008), while after ischemia carotid, ET-1 produced by endothelial cells is responsible for inflammation by increasing expression of adhesion molecules and formation of neointimal (Anggrahini et al., 2009).

In endothelial cells, ET-1 is stored in Weibel-Palade bodies (Rondaij et al., 2006). In cultured endothelial cells, ET-1 is released continuously through a constitutive pathway and is mainly regulated by gene transcription. When ocurs endothelial activation of cells by chemical stimuli, or mechanical (shear stress), the stored granules relocate from the cytoplasm to the plasma membrane and are released by exocytosis (Lowenstein et al., 2005). ET-1 then acting as in an autocrine or paracrine manner at the endothelin receptors present in endothelial cells (ETB) and smooth muscle cells (ETA and ETB) (Luscher and Barton, 2000).

In endothelial cells ET-1 mRNA expression is regulated by different biological or pharmacological factors. Transformed growth factor (TGF), thrombin, bradykinin, angiotensin II, the tumoral necrosis factor (TNF), glucocorticoids and hypoxia increases mRNA levels of ET-1, while vasodilators such as natriuretic peptide, nitric oxide, prostacyclin and prostaglandin, reduce the levels (Vignon-Zellweger et al., 2012).

Nitric oxide (NO) is one of the most widespread signaling molecules and participates in the functions of virtually all cells and organs of the body. The physiological levels of NO produced by the endothelial cells is essential to in adjusting the relaxation and proliferation of smooth muscle cells, leukocyte adhesion, platelet aggregation, thrombosis and angiogenesis (Ignarro et al., 1999). According to the mechanism described above, ET-1 should by autocrine and paracrine action, binding to ETB receptors on endothelial cells and leading to their activation and increased production of prostaglandin E2 (PGE2) and nitric oxide; these in turn should reduce the production of ET-1.

All compounds studied stimulated endothelial cells, causing increased production of ET-1; but only BSE presented increased nitric oxide and prostaglandin E2 production and reduced ET-1 production compared to the extract without biotransformation. However, we cannot to state that there was reduction in production of ET-1 induced by nitric oxide and prostaglandin E2, because for that, the production of ET-1 should be evaluated in several times, which did not fit the goals of the study. Further studies are needed to confirm this mechanism of action of BSE. For the other compounds, even with increased ET-1, we found no increase in the production of NO and PGE2 (De Andrade et al., 2014B).

White et al. (2010) suggest an interesting mechanism by which this difference might be explained. According to these researchers, as they tried to block the relaxation induced by estrogen coronary arteries which removed the endothelium by inhibiting the binding of L-arginine to NOS-1 encountered a strong coronary constriction induced by estrogen (White et al., 2005). These constrictor effects of estrogen could be reversed reducing the production of superoxide with tempol (one of superoxide dismutase mimic) or tiron one (superoxide scavenger), suggesting that vasoconstriction is mediated by reactive oxygen species. These researchers confirmed superoxide production stimulated by estrogen in smooth muscle cells where NO production was inhibited by blocking the L-arginine NOS-1 (White et al., 2005). They also demonstrated that the estrogen effect was completely dependent on the presence of L-arginine, showing a vasodilator effect in the presence of this cofactor and constrictive effect when its connection to NOS-1 was inhibited.

As the NOS is a multifunctional enzyme, possessing oxygenase and reductase domains with binding sites for several cofactors, the disposition of the microenvironment around the NOS is what will determine the primary product, generating superoxide instead of NO. Among the NOS isoform, NOS-1 is the one that exhibits greater tendency to produce superoxide under conditions where there is absence of substrate and cofactors (Alderton et al., 2001) (Csiszar et al., 2002) (Gorren and Mayer, 1998). However, all isoforms of NOS have the ability to produce superoxide, eNOS being one potential mechanism for the production of superoxide in the endothelium stimulated by estrogen, which may also contribute to the deleterious effects of estrogen on cardiovascular function (White et al., 2010).

This hypothesis could explain why only the BSE led to the production of NO, but further studies are needed to understanding this mechanism and to confirm this hypothesis (De Andrade et al., 2014B).

The results of this study are relevant because neither 17β-estradiol (E2) nor other phytoestrogens tested caused increased production of nitric oxide or PGE2 (De Andrade et al., 20014B). Moreover, BSE was able to reduce oxidative stress induced by hydrogen peroxide and decreased the expression of adhesion molecules (VCAM-1, ICAM-1 and E-selectin) on HUVEC cells, being more effective than other compounds in reducing sICAM-1 levels (De Andrade et al., 2014A). In addition, we found no studies in the literature evaluating together the three mechanisms that contribute to atherosclerosis (expression of adhesion molecules, oxidative stress and nitric oxide production).

CONCLUSION

Thus, we can conclude that the higher content of aglycone forms in BSE presented higher estrogenic potential, being the only one able to increase the production of nitric oxide and prostaglandin E2 (De Andrade et al., 2014B). As BSE was able to reduces the expression of adhesion molecules and oxidative stress and increases the nitric oxide production, it could be a very promising compound for both, prevention and treatment of atherosclerosis (De Andrade et al., 2014A), (De Andrade et al., 2014B).

References

Adamson, P., Tighe, M., Pearson, J. D. Protein tyrosine kinase inhibitors act downstream of IL-1 alpha and LPS stimulated MAP-kinase phosphorylation to inhibit expression of E-selectin on human umbilical vein endothelial cells. *Cell Adhes. Commun.* 1996;3:511-25.

Alderton, W. K., Cooper, C. E., Knowles, R. G. Nitric oxide synthases: structure, function and inhibition. *Biochem. J.* 2001;357:593-615.

Amiri, F., Paradis, P., Reudelhuber, T. L., Schiffrin, E. L. Vascular inflammation in absence of blood pressure elevation in transgenic murine model overexpressing endothelin-1 in endothelial cells. *J. Hypertens.* 2008;26:1102-9.

Andozia, B. M., Vieira, C. S., Franceschini, S. A., Toloi, M. R. T., et al. Ethinylestradiol and estradiol have different effects on oxidative stress and nitric oxide synthesis in human endothelial cell cultures. *Fertility and Sterility*. 2010; 94(5): 1578-1582.

Anggrahini, D. W., Emoto, N., Nakayama, K., Widyantoro, B., Adiarto, S., Iwasa, N., et al. Vascular endothelial cell-derived endothelin-1 mediates vascular inflammation and neointima formation following blood flow cessation. *Cardiovasc. Res.* 2009;82: 143-51.

Antonelli, A., Bianchi, M., Crinelli, R., Gentilini, L., Magnani, M. Modulation of ICAM-1 Expression in ECV304 Cells by Macrophage-Released Cytokines. *Blood Cells, molecules and diseases*, California, v. 27, n. 6, p. 978-991, nov./dec. 2001.

Akiyama, T., Ishida, J., Nakagawa, S., Ogawara, H., Watanabe, S., Itoh, N., Shibuya, M., Fukami, Y. Genistein, a specific inhibitor of tyrosine-specific protein kinases. *Journal of Biological Chemistry*, Baltimore, v. 262, n. 12, p. 5592-5595, April 1987.

Albelda, S. M., Oliver, P. D., Romer, L. H., Buck, C. A. EndoCAM: a novel endothelial cell-cell adhesion molecule. *J. Cell. Biol.* 1990; 110:1227-1237.

Andrade, C. M. Phytoestrogens antiatherogenic effect on adhesion molecules expression in endothelial cells., 2209. 101f. *Thesis (Doctorate)*. Faculty of Pharmaceutical Sciences of Ribeirao Preto - University of Sao Paulo, Ribeirao Preto, 2009.

Bagnall, A. J., Kelland, N. F., Gulliver-Sloan, F., Davenport, A. P., Gray, G. A., Yanagisawa, M., et al. Deletion of endothelial cell endothelin B receptors does not affect blood pressure or sensitivity to salt. *Hypertension* 2006;48:286-93.

Bakhit, R. M., Klein, B. P., Essex-Sorlie, D., Ham, J. O., Erdman, J. W. Jr, Potter, S. M. Intake of 25g of soybean protein with or without soybean fiber alters plasm lipids in men with elevated cholesterol concentrations. *Journal of Nutrition, Philadelphia*, v. 124, n. 2, p. 213-22. February 1994.

Barrett-Connor, E., Bush, T. L. Estrogen and coronary heart disease in women. *JAMA: the journal of the American Medical Association*, Chicago, v. 265, p. 1861-1867, Apr. 1991.

Beck, V., Rohr, U., Jungbauer, A. Phytoestrogens derived from red clover: an alternative to estrogen replacement therapy? *The Journal of Steroid Biochemistry and Molecular Biology*, Oxford, v. 94, p. 499-518, Mar. 2005.

Berliner, J. A., Navab, M., Fogelman, A. M., Frank, J. S., Demer, L. L., Edwards, P. A., Watson, A. D., Lusis, A. J. Atherosclerosis: basic mechanisms. Oxidation, inflammation, and genetics. *Circulation* 1995;91: 2488-96.

Bevilacqua, M. P., Pober, J. S., Mendrick, D. L., Cotran, R. S., Gimbrone, M. A. Jr. Identification of an inducible endothelial-leukocyte adhesion molecule. *Proceedings of the National Academy of Sciences*, US, Washington, v. 84, n. 24, p. 9238-9242, December 1987.

Birben, E., Sahiner, U. M., Sackesen, C., Erzurum, S., Kalayci, O. Oxidative stress and antioxidant defense. *World Allergy Organization Journal*, p. 9-19, 2012.

Blockhina, O., Virolainen, E., Fagerstedt, K. V. Antioxidants, oxidative damage and oxygen deprivation Stress: A review. *Ann. Bot.* 91: 179-184. 2003.

Bowie, A. G., Moynagh, P. N., O'Neill, L. A., Lipid peroxidation is involved in the activation of NF-κB by tumor necrosis factor but not interleukin-1 in the human endothelial cell line ECV304. Lack of involvement of H2O2 in NF-κB activation by either cytokine in both primary and transformed endothelial cells. *J. Biol. Chem.* 1997;272:25941-50.

Busconi, L., Michel, T. Endothelial nitric oxide synthase membrane targeting. Evidence against involvement of a specific myristate receptor. *The Journal of biological chemistry*, Baltimore, v. 7, p. 25016-2520, Oct. 1994.

Cano, A., Van Baal, W. M. The mechanisms of thrombotic risk induced by hormone replacement therapy. *Maturitas*, Limerick, v. 40, n. 1, p. 17-38, October 2001.

Carlos, T. M., Harlan, J. M. Leukocyte-endothelial adhesion molecules. *Blood*, New York, v. 84, n. 7, p. 2068-2101, October 1994.

Chambliss, K. L., Shaul, P. W. Estrogen modulation of endothelial nitric oxide synthase. *Endocrine Society*, Baltimore, v. 23, p. 665-686, Oct. 2002.

Chien, H. L., Huang, H. Y. and Chou, C. C., Transformation of isoflavone phytoestrogens during the fermentation of soymilk with lactic acid bacteria and bifidobacteria. *Food Microbiol.*, v. 23, p. 772-778, 2006.

Cho, H. Y., Park, C. M., Kim, M. J., Chinzorig, R., Won, C., Song, Y. S. Comparative effect of genistein and daidzein on the expression of MCP-1, eNOS, and cell adhesion molecules in TNF-α-stimulated HUVECs. *Nutrition Research and Practice (Nutr. Res. Pract.)* 2011;5(5):381-388.

Chung, J., Kim, S. Y., Jo, H. H., Hwang, S. J., Chae, B., Kwonb, D. J., et al. Antioxidant effects of equol on bovine aortic endothelial cells. *Biochemical and Biophysical Research Communications*. 375: 420-424. 2008.

Colacurci, N., et al. Effects of soy isoflavones on endothelial function in healthy postmenopausal women. *Menopause*, New York, v. 12, n. 3, p. 299-307, may/jn. 2005.

Collins, R. G., Velji, R., Guevara, N. V., Hicks, M. J., Chan, L., Beaudet, A. L. Pselectins or intercellular adhesion molecule (ICAM-1) deficiency substantially protects against atherosclerosis in apolipoprotein E-deficient mice. *Journal of Experimental Medicine*, New York, v. 191, n. 1, p. 189-194, January 2000.

Cos, P., et al. Phytoestrogens: recent developments. *Planta Medica*, Stuttgart, v. 69, n. 7, p. 589-599, July 2003.

Csiszar, A., Ungvari, Z., Edwards, J. G., Kaminski, P., Wolin, M. S., Koller, A., et al. Aging induced phenotypic changes and oxidative stress impair coronary arteriolar function. *Circ. Res.* 2002;90:1159-66.

De Andrade, C. M., Silva De Sá, M. F., Toloi, M. R. T. Effects of phytoestrogens derived from soy bean on expression of adhesion molecules on HUVEC. *Climacteric*. April 2012, Vol. 15, No. 2: Pages 186-194.

De Andrade, C. M., Bianchini, F. J., Rey, F. M., Fonseca, M. J. V., Toloi, M. R. T. Effects of an aglycone-rich biotransformed soybean extract on the expression of adhesion molecules and oxidative stress induced by hydrogen peroxide in human endothelial cells. *Gior. It. Ost. Gin.* 2014;36 (1):111-116. (A).

De Andrade, C. M., Bianchini, F. J., Rey, F. M., Fonseca, M. J. V., Toloi, M. R. T. Effects of an aglycone-rich biotransformed soybean extract in human endothelial cells. *Climacteric*. 2014; 00:1-5. (B).

De Caterina, R., Libby, P., Peng, H. B., Thannnickall, V. J., Rajavashisth, T. B., Gimbrone, M. A. Jr, Shin, W. S., Liao, J. K. Nitric oxide decreases cytokine-induced endothelial activation: nitric oxide selectively reduces endothelial expression of adhesion molecules and proinflammatory cytokines. *J. Clin. Invest.*, 96:60-68, 1995.

Dustin, M. L., Springer, T. A. Lymphocyte function associated antigen (LFA-1) interacting with intercellular adhesion molecule-1 (ICAM-1) is one of at least three mechanisms for lymphocyte adhesion to culture endothelial cells. *Journal of Cell Biology*, New York, v. 107, n. 1, p. 321-331, July 1988.

Eden, J. A. Managing the menopause: phytoestrogens or hormone replacement therapy? *Annals of Medicine*, Helsinki, v. 33, n. 1, p. 4-6, February 2001.

Endemann, D., Schiffrin, E. Endothelial Dysfunction. *The American Society of Nephrology*, Washington, v. 15, p. 1983-1922, Aug. 2004.

Fujio, Y., Yamada, F., Takahashi, K., Shibata, N. Responses of smooth muscle cells to platelet-derived growth factor are inhibited by herbimycin-A tyrosine kinase inhibitor. *Biochemical and Biophysical Research Communications*, New York, v. 195, n. 1, p. 79-83, August 1993.

Galli, F., Piroddi, M., Annetti, C., Aisa, C., Floridi, E., Floridi, A. Oxidative stress and reactive oxygen species. *Contrib. Nephrol.* 149:240-269. 2005.

Georgetti, S. R., Vicentini, F. T. M. C., Yokoyama, C. Y., Borin, Spadaro, A. C. C., Fonseca, M. J. V. Enhanced *in vivo* and *in vitro* antioxidant activity and mobilization of free phenolic compounds of soybean flour fermented with different β-glucosidaseproducing fungi. *Journal of Applied Microbiology*, v. 106, p. 459-466, 2009.

Gong, G., Qin, Y., Huang, W., Zhou, S., Yang, X., Li, D. Rutin inhibits hydrogen peroxide-induced apoptosis through regulating reactive oxygen species mediated mitochondrial dysfunction pathway in human umbilical vein endothelial cells. *European Journal of Pharmacology*, v. 628, p. 27-35. 2010.

Gorren, A. C., Mayer, B. The versatile and complex enzymology of nitric oxide synthase. *Biochemistry* (Mosc.) 1998;63:734-43.

Grodstein, F., et al. Postmenopausal estrogen and progestin use and the risk of cardiovascular disease. *New England Journal of Medicine*, Boston, v. 335, n. 7, p. 453-461, August 1996.

Gueteens, G. G., De Boeck, M., Highley, A. T. Oesterom and E. A. Oxidative DNA damage: Biological significance and methods of analysis. *Crit. Rev. Clin. Lab. Sci.*, 39:331-457. 2002.

Haller, H. Endothelial function. General considerations. *Drugs*, New York, v. 53, p. 1-10, Jan. 1997.

Henderson, V. W., et al. Estrogen reolacement therapy in older women: comparisons between Alzheimer's disease cases and nondemented control subjects. *Archieves of Neurology*, Chicago, v. 51, n. 9, p. 896-900, September 1994.

Hewett, P., Murray, J. C. Human lung microvessel endothelial cells: isolation, culture and characterization. *Microvasc. Res.* 1993; 46:89-102.

Hewett, P. W., Murray, J. C. Coexpression of flt-1, flt-4 and KDR in freshly isolated and cultured human endothelial cells. *Biochem. Biophy. Res. Commun.* 1996; 221:697-702.

Humphrey, L. L., Chan, B. K. S., Sox, H. C. Postmenopausal Hormone Replacement Therapy and the Primary Prevention of Cardiovascular Disease. *Annals of Internal Medicine*, Philadelphia, v. 137, n. 4, p. 273-284, August 2002.

http://www.anvisa.gov.br/legis/resol/2003/re/899_03re.htm. Acesso em 10 de maio de 2010.

Huo, Y., Ley, K. Adhesion molecules and atherogenesis. *Acta Physiologica Scandinavica*, Stockholm, v. 173, n. 1, p. 35-43, September 2001.

Hurley, S., Grady, D., Bush, T., et al. Randomized trial of estrogen plus progestin for secondary prevention of coronary heart disease in postmenopausal women. Heart and Estrogen/progestin Replacement Study (HERS). *JAMA: The Journal of the American Medical Association*, Chicago, v. 280, p. 605-613. Aug. 1998.

Ignarro, L. J., Cirino, G., Casini, A., Napoli, C. Nitric oxide as a signalling molecule in the vascular system: an overview. *J. Cardiovasc. Pharmacol.* 1999; 879-86.

Jaffe, E. A., Hoyer, L. W., Nachman, R. L. Synthesis of von Willebrand factor by cultured human endothelial cells. *Proc. Natl. Acad. Sci. US.* 1974; 1906-1909.

Jonsson, A. S., Palmblad, J. E. Effects of ethanol on NF-Kappa B activation , production of myeloid growth factors , and adhesive events in human endothelial cells. *Journal of Infectious Diseases*, Chicago, v. 184, n. 6, p. 761-769, September 2001.

Kansas, G. S. Selectins and their ligands: current concepts and controversies. *Blood*, New York, v. 88, n. 9, p. 3259-3287, November 1996.

Kelland, N. F., Bagnall, A. J., Morecroft, I., Gulliver-Sloan, F. H. Dempsie, Y., Nilsen, M., et al. Endothelial ETB limits vascular remodelling and development of pulmonary hypertension during hypoxia. *J. Vasc. Res.* 2010;47:16-22.

Kelland, N. F., Kuc, R. E., McLean, D. L., Azfer, A., Bagnall, A. J., Gray, G. A., et al. Endothelial cell-specific ETB receptor knockout: autoradiographic and histological characterisation and crucial role in the clearance of endothelin-1. *Can. J. Physiol. Pharmacol.* 2010;88:644-51.

Kuiper, G. G., Lemmen, J. G., Carlsson, B., Corton, J. C., Safe, S. H., van der Saag, P. T., van der Burg, B., Gustafsson, J. A. Interaction of estrogenic chemicals and phytoestrogens with estrogen receptor β. *Endocrinology*, Springfield, 1998, v. 139, p. 4252-4263, Oct. 1998.

Lee, H. C., Yang, L., Xu, J. Z., Yeung, S. Y. V., Huang, Y., Chen, Z. Y., Relative antioxidant activity of soybean isoflavones and their glycosides. *Food Chem.*, v. 90, p. 735-741, 2005.

Leikert, J. F., Rãthel, T. R., Vollmar, A. M., Dirsch, V. M. Reliable in vitro measurement of nitric oxide released from endothelial cells using low concentrations of thefluorescent probe 4,5-diaminofluorescein. *FEBS letters,* Amsterdam, v. 5, p. 131-134, Oct. 2001.

Ley, K. The role of selectins in inflammation and disease. *Trends in Molecular Medicine*, Oxford, v. 9, n. 6, p. 263-268, June 2003.

Libby, P., Ordovas, J. M., Auger, K. R., Robbins, A. H., Birinyi, L. K., Dinarello, C. A. Endotoxin and tumor necrosis factor induce inteleukin-1 gene expression in adult human vascular endothelial cells. *American Journal of Pathology*, New York, v. 124, n. 2, p. 179-185, August 1986.

Lin, C. H., Wei, Y. T. and Chou, C. C. Enhanced antioxidative activity of soybean koji prepared with various filamentous fungi. *Food Microbiol.*, v. 23, p. 628-633, 2006.

Liu, D., Homan, L. L., Dillon, J. S. Genistein acutely stimulates nitric oxide synthesis in vascular endothelial cell by a cyclic adenosine 5-monophosphate-dependent mechanism. *Endocrinology*, Springfield, v. 12, p. 5532-5539, Dec. 2004.

Lowenstein, C. J., Morrell, C. N., Yamakuchi, M. Regulation of Weibel-Palade body exocytosis. *Trends Cardiovasc. Med.* 2005;15:302-8.

Luscher, T. F., Barton, M. Endothelins and endothelin receptor antagonists: therapeutic considerations for a novel class of cardiovascular drugs. *Circulation* 2000;102:2434-40.

Lteif, A. A., Han, K., Mather, K. J. Obesity, insulin resistance and the metabolic syndrome: determinants of endothelial dysfunction in whites and blacks. *Circulation*, Dallas, v. 112, n. 1, p. 32-38, July 2005.

Mahmoudi, M., Curzen, N., Gallagher, P. J. Atherosclerosis: the role of inflammation and infection. *Histopathology*, Oxford, v. 50, p. 534-546, April. 2007.

Marasciulo, F. L., Montagnani, M., Potenza, M. A. Endothelin-1: the yin and yang on vascular function. *Current medicinal chemistr.*, Schiphol., v. 13, p. 1655-1665, Mai. 2006.

Matti, J., Adlercreutz, T., Adlercreutz, H. Dietary Soy-Derived Isoflavone Phytoestrogens. Could they have a role in coronary heart disease prevention? *Biochemical Pharmacology*, Oxford, v. 60, n. 1, p. 1-5, July 2000.

Maturana, M. M. A., Irigoyen, M. C., Spritzer, P. M. Menopause, estrogens, and endothelial dysfunction. *Clinics: official scientific journal of Hospital das Clínicas*, Faculty of Medicine, University of São Paulo, São Paulo, v. 62, p. 77-86, Feb. 2007.

May, M. J., Wheeler-Jones, C. P., Pearson, J. D. Effects of protein tyrosine kinase inhibitors on cytokine-induced adhesion molecule expression by human umbilical vein endothelial cells. *Br. J. Pharmacol.* 1996;118:1761-71.

Mei, J., Yeung, S. S. C., Kung, A. W. C. High dietary phytoestrogen intake is associated with higher bone mineral density in postmenopausal but not premenopausal women. *Journal of Clinical Endocrinology and Metabolism*, Philadelphia, v. 86, n. 11, p. 5217-5221, November 2001.

Mendelsohn, M. E., Karas, R. H. Molecular and cellular basis of cardiovascular gender differences. *Science*, Washington, v. 308, p. 1583-1587, Jun. 2005.

Merrit, J. C. Metabolic Syndrome: Soybean Foods and Serum Lipids. *Journal of National Medical Association*, New York, v. 96, n. 8, p. 1032-1042, August 2004.

Mohammad, S. F. and Bos, W. J. W. - Physiology of the vessel blood interface. In: Bunt, T. J. (ed.) - latrogenic vascular injury. *A Discourse an surgical technique.* New York, Futura Publishing, p. L - 14, 1990.

Moreno, J. J. Antiflammin-2 Prevents HL-60 Adhesion to Endothelial Cells and Prostanoid Production Induced by Lipopolysaccharides. *The Journal of Pharmacology and Experimental Therapeutics*, Baltimore, v. 296, n. 3, p. 884-889, March 2001.

Mosmann, T. Rapid colorimetric assay for cellular growth and survival: application to proliferation and cytotoxicity assays. *Journal of immunological methods*, Amsterdam, v. 16, p. 55-56, Dec. 1983.

Muller, W. A. Leukocyte-endothelial cell interactions in the inflammatory response. *Laboratory Investigation*, Baltimore, v. 82, n. 5, p. 521-533, May 2002.

Naessen, T., Persson, I., Adami, H. O. Hormone replacement therapy and the risk for first hip fracture. A prospective population-based cohort study. *Annals of Internal Medicine*, Philadelphia, v. 113, n. 2, p. 95-103, July 1990.

Nagata, C., Takatsuka, N., Shimizu, H. Soy and fish oil intake and mortality in a Japanese community. *American Journal of Epidemiology*, Baltimore, v. 156, n. 9, p. 824-831, November 2002.

Nageh, M. F., Sandberg, E. T., Marotti, K. R., et al. Deficiency of inflammatory cell adhesion molecules protects against atherosclerosis in mice. *Arterioscler. Thromb. Vasc. Biol.* 1997;17:1517e1520.

Nestel, P. J., Yamashita, T., Sasahara, T., et al. Soy isoflavones improve systemic arterial compliance but not plasma lipids in menopausal and perimenopausal women. *Arteriosclerosis, Thrombosis vascular Biology*, Dallas, v. 17, n. 12, p. 3392-3398, December 1997.

Nicholls, D. G. and Budd, S. L. Mitochondria and neuronal survival. *Physiol. Rev.*, 80:315-360. 2000.

Nohl, H., Kozlov, A. V., Gille, L., Staniek, K., Cell respiration and formation of reactive oxygen species: facts and artefacts. *Biochem. Soc. Trans.* 31, 1308-1311. 2003.

Noubir, S., Hmana, Z., Reiner, N. E. Dual receptors and distinct pathways mediate IRAK degradation in response to lipopolysaccharide: involvement of CD14/TLR4, CR3 and phosphatidylinositol 3-kinase, *Journal of Biological Chemistry*, Baltimore, v. 279, n. 24, p. 25189-25195, June 2004.

Osborn, L., Hession, C., Tizard, R., Vassallo, C., Luhowskyj, S., Chi-Rosso, G., Lobb, R. Direct expression cloning of vascular cell adhesion molecule 1, a cytokine-induced endothelial protein that binds to lymphocytes. *Cell*, Cambridge, v. 59, n. 6, p. 1203-1211, December 1989.

Ososki, A. L., Kennely, E. J. Phytoestrogens as cardioprotective agents. *Phytotherapy Research: PTR.*, v. 17(3), p. 845-869, 2003.

Palmer, R. M., Ferrige, A. G., Moncada, S. NO release accounts for the biological activity of endothelium-derived relaxing factor. *Nature*, London, v. 327, p. 524-526. Sep. 1987.

Parkington, H. C., Coleman, H. A., Tare, M. Prostacyclin and endothelium-dependent hyperpolarization. *Pharmacological research: the official journal of the Italian Pharmacological Society*, London, v. 49, p. 509-514, Jun. 2004.

Paulo, M., Salvador, M. M., Dos Anjos Neto Filho, M., Montes, M. B., Franceschini, S. A., Toloi, M. R. Effect of isoflavone extracts from Glycine max on human endothelial cell damage and on nitric oxide production. *Menopause*. May-Jun; 16(3):539-44, 2009.

Potdara, S., Kavdia, M. NO/peroxynitrite dynamics of high glucose-exposed HUVECs: chemiluminescent measurement and computational model. *Biol. Chem.* 390, 899-905. 2009.

Quehenberger, P., Bierhaus, A., Fasching, P., Muellner, C., Klevesath, M., Hong, M., et al. Endothelin 1 transcription is controlled by nuclear factorkappa-b in age-stimulated cultured endothelial cells. *Diabetes* 2000; 49: 1561-70.

Quella, S. K., Loprinzi, C. L., Barton, D. L., Knost, J. A., Sloan, J. A., Lavasser, B. I., Swan, D., Krupp, K. R., Miller, K. D., Novotny, P. J. Evaluation of phytoestrogens for the treatment of hot flashes in breast cancer survivors: a North Central Cancer Treatment Group Trial. *Journal of Clinical Oncology* (New York), v. 18(5), p. 1068-74, 2000.

Ray, R. S., Mehrotra, S., Shankar, U., Babu, G. S., Joshi, P. C., Hans, R. K. Evaluation of UV-induced superoxide radical generation potential of some commom antibotics. *Drugs Chem. Toxicol.*, 24; 191-200. 2001.

Rimbach, G., Boesch-Saadatmandi, C., Frank, J., Fuchs, D., Wenzel, U., Daniel, H., Hall, W. L., Weinberg, P. D. Dietary isoflavones in the prevention of cardiovascular disease--a molecular perspective. *Food Chem. Toxicol.* 2008;46:1308-19.

Rios, D. R., Rodrigues, E. T., Cardoso, A. P., Montes, M. B., Franceschini, S. A., Toloi, M. R. Effects of isoflavones on the coagulation and fibrinolytic system of postmenopausal women. *Nutrition*, Los Angeles, v. 24, p. 120-126, Jan. 2008.

Roberts, J. M. and Hubel, C. A. Oxidative stress in preeclampsia. *Am. J. obstet. Gynecol*, v. 190, p: 1177-1178, 2004.

Rondaij, M. G., Bierings, R., Kragt, A., van Mourik, J. A., Voorberg, J. Dynamics and plasticity of weibel-palade bodies in endothelial cells. *Arterioscler. Thromb. Vasc. Biol.* 2006;26:1002-7.

Rymmer, J., Morris, E. P. Extracts from Clinical Evidence: Menopausal symptoms. *BMJ*, London, v. 321, n. 7275, p. 1516-1519, December 2000.

Schneider, M. P., Boesen, E. I., Pollock, D. M. Contrasting actions of endothelin ET(A) and ET(B) receptors in cardiovascular disease. *Annual review of pharmacology and toxicology*, Palo Alto, v. 47, p. 731-759, Fev. 2007.

Schutz, M. Friedl, P. Isolation and cultivation of endothelial cells derived from human placenta. *Eur. J. Cell Biol.* 1996; 71: 395-401.

Sellman, S. Osteoporosis: the bones of contention. *Nexus New Time Magazine*, v. 5, n. 6, p. 1-8. 1998.

Shimokado, K., Yokota, T., Umezawa, K., Sasagurli, T., Ogata, J. Protein tyrosine kinase inhibtors inhibit chemotaxis of vascular smooth muscle cells. *Ateriosclerosis and Thrombosis*, Dallas, v. 14, n. 6, p. 973-981, June 1994.

Simoncini, T., et al., Activation of nitric oxide synthesis in human endothelial cells by red clover extracts. *Menopause*, New York, v. 12 n. 1 p. 69-77 jan/feb 2005.

Simoncini, T., et al. Effects of phytoestrogens derived from red clover on atherogenic adhesion molecules in human endothelial cells. *Menopause*, New York, v. 15 n. 3, p. 542-550. 2008.

Stankevicius, E., Kevelaitis, E., Vainorius, E., Simonsen, U. Role of nitric oxide and other endothelium-derived factors. *Medicina* (Kaunas, Lithuania), Lithuania, v. 39, p. 333-341, Jun. 2003.

Störk, S., Von Schacky, C., Angerer, P. The effect of 17B-estradiol on endothelial and inflammatory markers in postmenopausal women: a randomized controlled trial. *Atherosclerosis*, Amsterdam, v. 165, n. 2, p. 301-307, December 2002.

Tang, M. X., et al. Effect of oestrogen during menopause on risk and age at onset of Alzheimer´s disease. *Lancet*, London, v. 348, n. 9025, p. 429-432, August 1996.

Tedder, T. F., Steeber, D. A., Chen, A., Engel, P. The selectins: vascular adhesion molecules. *FASEB Journal*, Bethesda, v. 9, n 10, p. 866-873, July 1995.

Thorin, E., Webb, D. J. Endothelium-derived endothelin-1. Pflugers *Arch.* 2010;459:951-8.

Tirapelli, C. R., Casolari, D. A., Yogi, A., Montezano, A. C., Tostes, R. C., Legros, E., et al. Functional characterization and expression of endothelin receptors in rat carotid artery: involvement of nitric oxide, a vasodilator prostanoid and the opening of K+ channels in ETB-induced relaxation. *Br. J. Pharmacol.* 2005;146:903-12.

Tsukahara, H., Ende, H., Magazine, H. I., Bahou, W. F., Goligorsky, M. S. Molecular and functional characterization of the non-isopeptide-selective ETB receptor in endothelial cells. Receptor coupling to nitric oxide synthase. *J. Biol. Chem.* 1994;269:21778-85.

Unger, R. E., Krump-Konvalinkova, V., Peters, K., Kirkpatrick, C. J. In vitro expression of endothelial phenotype: comparative study of primary isolated cells and cell lines, including the novel cell line HPMEC-ST1.6R. *Microvascular Research*. 2002; 64: 384-397.

Vera, R., Sánchez, M., Galisteo, M., Villar, I. C., Jimenez, R., Zarzuelo, A., Pérez-Vizcaíno, F., Duarte, J. Chronic administration of genistein improves endothelial dysfunction in spontaneously hypertensive rats: involvement of eNOS, caveolin and calmodulin expression and NADPH oxidase activity. *Clin. Sci.* (Lond.) 2007;112:183-91.

Verbon, A., et al. IC14, an anti-CD14 antibody, inhibits endotoxin-mediated symptoms and inflammatory responses in humans, *Journal of Immunology*, Baltimore, v. 166, n. 5, p. 3599-3605, March 2001.

Verhoeven, M. O., Van Der Mooren, M. J., Van De Weijer, P. H., Verdegem, P. J., Van Der Burgt, L. M., Kenemans, P.; Curatrial Research Group. Effect of a combination of isoflavones and Actaea racemosa Linnaeus on climacteric symotons in healthy symptomatic perimenopausal women: a 12-week randomized, placebo-controlled, double–blind study. *Menopause*, New York, v. 12, n. 4, p. 412-420, jul/aug. 2005.

Vignon-Zellweger, N., Heiden, S., Miyauchi, T., Emoto, N. Endothelin and endothelin receptors in the renal and cardiovascular systems. *Life Sciences* 2012; 91: 490-500.

Xu, X., et al. Bioavailability of soybean isoflavones depends upon gut microflora in women. *Journal of Nutrition*, Philadelphia, v. 125, n. 9, p. 2307-2317, September 1995

Wagner, O. F., Christ, G., Wojta, J., Vierhapper, H., Parzer, S., Nowotny, P. J., et al. Polar secretion of endothelin-1 by cultured endothelial cells. *J. Biol. Chem.* 1992;267:16066-8.

Wang, H. J. and Murphy, P. A., Isoflavone content in commercial soybean foods, *Journal Agricultural and Food Chemistry*, v. 42, p. 1666-1673, 1994.

Wang, H., Zhao, X., Yin, S. Effects of coenzyme Q10 or combined with micronutrients on antioxidant defense system in rats. *Wei Sheng Yan Jiu*. 37:311-313. 2008.

Washburn, S., Burke, G. L., Morgan, T., Anthony, M. Effect of soy protein supplementation on serum proteins, blood pressure, and menopausal symptons in perimenopausal women. *Menoipause*, New York, v. 6, n. 1, p. 7-13, 1999.

Webb, D. J., Haynes, W. G. The role of endothelin-1 in cardiovascular physiology and pathophysiology. *Scottish medical journal*, Glasgow, v. 40, p. 69-71, Jun. 1995.

Weiss, N. S., Ure, C. L., Ballard, J. H. Decreased risk of fractures of the hip and lower forearm with postmenopausal use of estrogen. *New England Journal of Medicine*, Boston, v. 303, n. 21, p. 1195-1198, November 1980.

Wenzel, U., Fuchs, D., Daniel, H. Protective effects of soy-isoflavones in cardiovascular disease. Identification of molecular targets. *Hamostaseologie* 2008;28:85e88.

WHI- The Writing Group for the Women's Health Initiative Investigators. Risks and benefits of estrogen plus progestin in healthy postmenopausal women: principal results from the Women's Health Initiative randomized controlled trial. *JAMA: The journal of the American Medical Association*. Chicago, v. 288, p. 321-333, Jul. 2002.

White, R. E., Gerrity, R., Barman, S. A., Han, G. Estrogen and oxidative stress: A novel mechanism that may increase the risk for cardiovascular disease in women. *Steroids* 2010; 75: 788-793. (B).

White, R. E., Hang, G., Dimitropoulou, C., Zhu, S., Miyake, K., Fulton, D., et al. Estrogen-induced contraction of coronary arteries is mediated by superoxide generated in vascular smooth muscle. *Am. J. Physiol. Heart Circ. Physiol.* 2005; 289:H1468-75. (A).

Wolin, M. S. Reactive oxygen species and vascular signal transduction mechanisms. *Microcirculation*, New York, v. 3, p. 1-17, Mar. 1996.

Wolk, K., Kunz, S., Crompton, N. E., Volk, H. D., Sabat, R. Multiple mechanisms of reduced major histocompatibility complex class II expression in endotoxin tolerance, *Journal of Biological Chemistry*, Baltimore, v. 278, n. 20, p. 18030-18036, May 2003.

Wuttke, W., et al. Phytoestrogens for hormone replacement therapy? *Journal of Steroid Biochemistry and Molecular Biology*, Oxford, v. 83, n. 1-5, p. 133-147, December 2002.

Yaffe, K., et al. Estrogen therapy in postmenopausal women: effects on cognitive function and dementia. *JAMA*, Chicago, v. 279, n. 9, p. 688-695, March 1998.

Yanagisawa, M., Kurihara, H., Kimura, S., Tomobe, Y., Kobayashi, M., Mitsui, Y., et al. A novel potent vasoconstrictor peptide produced by vascular endothelial cells. *Nature* 1988;332:411-5.

Yang, B., Oo, T. N., Rizzo, V. Lipid rafts mediate H2O2 prosurvival effects in cultured endothelial cells. *FASEB J.* 20, 1501-1503. 2006.

Zakynthinos, E., Pappa, N. Inflammatory biomarkers in coronary artery disease. *J. Cardiol.* 2009;53:317-333.

Zhu, H., Zhang, L., Amin, A. R., Li, Y. Coordinated upregulation of a series of endogenous antioxidants and phase 2 enzymes as a novel strategy for protecting renal tubular cells from oxidative and electrophilic stress. *Exp. Biol. Med.* (Maywood), 233:753-765. 2008.

In: Genistein and Daidzein
Editor: Masayoshi Yamaguchi
ISBN: 978-1-63483-194-9

Chapter 3

ROLE OF GENISTEIN IN BONE HOMEOSTASIS: PREVENTION OF OSTEOPOROSIS

Masayoshi Yamaguchi *
Department of Hematology and Medical Oncology,
Emory University School of Medicine, Atlanta, US

ABSTRACT

Bone homeostasis is regulated through osteoclasts and osteoblasts. Osteoporosis, which is induced with its accompanying decrease in bone mass with increasing age, is widely recognized as a major public health problem. Bone loss may be due to decreased osteoblastic bone formation and increased osteoclastic bone resorption. There is growing evidence that nutritional and food factors may play a part in the prevention of bone loss with aging and have been to be worthy of notice in the prevention of osteoporsis. Genistein, which is contained in soybeans, has been shown to have a stimulative effect on osteoblastic bone formation and a suppressive effect on osteoclastic bone resorption, thereby increasing bone mass. This factor has a regulative effect on protein synthesis and gene expression, which are related to bone formation in osteoblastic cells and bone resorption in osteoclastic cells. Oral administration of genistein has been shown to prevent on bone loss in ovariectomized rats, an animal model for osteoporosis. Moreover, intake of dietary soybean genistein has been shown to have a preventive effect on osteoporosis in human subjects. These findings have been demonstrated a role of genistein in the prevention of bone loss with aging. This chapter introduces a role of dietary soybean genistein in osteoporosis prevention.

Keywords: Genistein, isoflavone, bone, osteoporosis, osteoblastic bone formation, osteoclastic bone resorption

* Corresponding author: Masayoshi Yamaguchi, Ph.D., Department of Hematology and Medical Oncology, Emory University School of Medicine, 1365 C Clifton Road, NE, Atlanta, GA 30322, US. E-mail: yamamasa1155@yahoo.co.jp.

INTRODUCTION

Bone homeostasis is skillfully regulated through osteoclasts and osteoblasts. Aging induces a reduction in bone mass by decrease in osteoblactic bone formation and increase in osteoclastic bone resorption. Osteoporosis with its accompanying decrease in bone mass is widely recognized as a major public health problem. The most dramatic expression of the disease is represented by fractures of the proximately femur, and the number of which increases as the population ages [1-3].

Malnutrition or undernutrition is often observed in the elderly, and it appears to be more intense in patients with hip fracture than in the general aging population [4]. Deficiency in both micronutrients and macronutrients appears to be strongly implicated in the pathogenesis and the consequences of hip fracture in the osteoporotic elderly. There is growing evidence that nutritional factors have a role in the prevention of bone loss with increasing age [5-7]. Food life may help to prevent bone loss with aging. Chemical compounds in food and plants, which regulate bone homeostasis, have been to be notice.

Recent studies have been shown that nutritional and functional food factors, which are present in fruit and vegetables, stimulate osteoblastic bone formation and suppress osteoclastic bone resorption in osteoporosis animal models and human subjects [5-7], thereby increasing bone mass. Supplemental intake of ingredients plays a role in maintaining of bone health and in the prevention of bone loss.

This chapter introduces recent findings concerning the role of soybean genistein in the regulation of bone homeostasis and in the prevention of osteoporosis in animal models and human subjects. Moreover, interesting topics have been focused that the anabolic effects of genistein on bone are enhanced by the combination with zinc, an essential trace element.

GENISTEIN STIMULATES BONE FORMATION

Bone Homeostasis

Bone homeostasis is regulated through the functions of osteoblasts and osteoclasts, which are major cells in bone tissues [8-14]. Osteoblasts, which are differentiated from bone marrow messenchymal stem cells, stimulate bone formation and calcification, while osteoclasts, which develop from hematopoietic progenitors, promote bone resorption.

Osteoclasts are recruited to the site and excavate the calcified matrix. In the physiologic process of bone turnover, a resorptive stimulus firstly triggers recruitment of osteoclasts to a site on the bone surface. After the resorbed lacunar pit is filled with new osteoid, osteoblasts become flatter and less active, with the final newly remodeled bone surface lined by flat lining cells. Remodeling of cancellous bone begins with the retraction of lining cells that cover the bone surface [13].

Bone acts as major storage site for growth factors, which are produced by osteoblasts, diffuse into newly deposited osteoid and are stored in the bone matrix including isulin-like growth factors (IGF- I and II), transforming growth factor-β1 (TGF-β1), platelet-derived growth factor (PDGF), or bone morphologic proteins (BMPs) [12, 14].

These bone-derived factors, which can be liberated during subsequent periods of bone resorption, act in an autocrine, paracrine, or delayed paracrine fashion in the local microenvironment of the bone surface.

Process of bone remodeling to make bone is unique among organs and tissues, and it is complexity with respect to interactions along the remodeling sequence by systemic influences (hormones), stress action on trabecular and cortical systems, growth factors and cytokines produced by the bone cells which act locally on their own cell types and on the other bone cell types, or factors that come from nearby cells present in the bone marrow tissues. Genistein has been found to possess a regulatory effect in the process of bone remodeling.

Genistein Has an Anabolic Effect in Bone Tissues

Isoflavones (including daidzin, daidzein, genistein, and genistein) are contained in soybeans with relatively high concentrations. Daidzin or genistin is hydrolyzed to daidzein or genistein by β-glucosidase in the gastrointestinal system, respectively. Genistein has been found to have a strong inhibitory effect on protein tyrosine kinase and it can cause cell cycle arrest and apotosis in leukemic cells [15]. This cellular mechanism of genistein may be important in the prevention of cancers. Yamaguchi et al. found that genistein or daidzein has been demonstrated to have a direct anabolic effect on bone metabolism *in vitro* [16], suggesting a role in the prevention of osteoporosis. This was the first time finding. Genistein as a food factor is a usefulness in the prevention and treatment of osteoporosis.

Anabolic effects of genistein on bone metabolism have been demonstrated in tissue culture using the femoral metaphyseal (trabecular bone) tissues obtained from elderly female rats *in vitro* [16, 17]. Bone tissues were cultured for 24 hours in a Dulbecco's modified Eagle's medium (high glucose, 4.5%) with bovine serum albumin (serum free) containing either vehicle or genistein. The presence of genistein (10^{-6} or 10^{-5} M) caused a significant increase in bone components (calcium content, alkaline phosphatase activity, and DNA content) in rat femoral metaphyseal tissues [16, 17]. The effect of genistein in increasing bone components is equal to the stimulatory effects of 17β-estradiol, which is important anabolic hormone in bone metabolism. Anti-estrogen agent tamoxifen inhibited the genistein-induced increase in bone components, although tamoxifen itself had no effect on bone components [16]. Anabolic effects of genistein on bone metabolism may be partly mediated through estrogen-like action. Presumably, genistein binds to the receptor of estrogen in osteoblastic cells in which the receptor is localized [18].

The effects of genistein or genistin on bone components have been compared in an *in vitro* culture system [16]. Culture with genistein (10^{-8} - 10^{-5} M) or genistin (10^{-7} - 10^{-5} M) increased alkaline phosphatase activity and deoxyribonucleic acid (DNA) and calcium contents in the femoral-metaphyseal tissues obtained from elderly female rats [16].

The effect of genistein was greater than that of genistin. The effects of genistein or genistin in increasing bone components were blocked in the presence of cycloheximide, an inhibitor of protein synthesis in the translational process, suggesting that the anabolic effects of the isoflavones are partly resulted from newly synthesized protein components.

Anabolic effects of daizein on the femoral-diaphyseal (cortical bone) tissues obtained from elderly female rats *in vitro* have also been shown [19]. The presence of daidzein (10^{-6} or 10^{-5} M) in culture medium caused a significant increase in bone components.

The effect of daizein was equal to that of genistein (10^{-6} or 10^{-5} M). The combination of daidzein and genistein did not have an additive effect [19]. In addition, the stimulating effect of daidzein on bone components was completely suppressed in the presence of cycloheximide [19], suggesting that this effect is resulted from protein synthsis.

The effects of phosphogenistein and phosphodaidzein, which are phosphorylated for the hydroxyl group at the 7-position of genistein and daidzein, on bone components has been compared to the effect of genistein or daizen [20]. Phosphogenistein or phosphodaidzein increased bone components in tissue culture *in vitro*. Phosphoisoflavones with a lower concentration of 10^{-6} M, at which genistein and daidzein have anabolic effects on bone components, did not have an effect [20].

The effects of various polyphenols found in food and plants on bone metabolism have also been compared [21]. Glycitein, resveratol, quercetin, catechin, and (-)-epigallocatechin gallate did not have an anabolic effect on bone calcification *in vitro* [21]. Polyphenols did not have a direct stimulative effect on bone calcification *in vitro*. Among various polyphenol, genistein has a unique anabolic effect on bone.

As the mentioned above, genistein is a 4', 5, 7-trihydroxyisoflavone, and daidzein is a 4', 7-dihydroxyisoflavone. Glycitein is a 4', 7-dihydroxy-6-methoxyflavone, while quercetin is a 3, 3', 4', 5, 7-pontahydeoxyflavone; neither has any effect of genistein and daidzein on bone calcification is weakened by the phosphorylation of the hydroxy-group at their-7-position [21]. The hydroxy-groups at the 5- and 7-positions of the isoflavone genistein may be necessary for its anabolic effect on bone formation, suggesting a chemical structure-activity relationship. Presumably, the chemical form of dihydroxyisoflavone is important in revealing its anabolic effects on bone. Genistein and daidzein may have bone effect that operates via the same mechanism.

Genistein Stimulates Osteoblastogenesis

Osteoblasts play an essential role in the stimulation of bone formation and calcification. The anabolic effect of isoflavones on osteoblastic cells *in vitro* has been demonstrated [22-24]. Osteoblastic MC3T3-E1 cells were cultured for 48 hours in a serum-free modified Eagle's medium containing genistein or daidzein. The presence of genistein (10^{-6} or 10^{-5} M) or daidzein (10^{-6} or 10^{-5} M) caused a significant increase in protein content, alkaline phosphatase activity and DNA content in the cells [22]. The ability of genistein or daidzein to increase biochemical components in the cells was not seen in the presence of an inhibitor of protein synthesis [22], suggesting that the effects of isoflavones result from newly synthesized protein components. The anabolic effects of genistein in osteoblastic cells are not distinguishable from those of daidzein [22, 23].

The effects of 17β-estradiol in increasing protein content and alkaline phosphatase activity in osteoblastic cells were not enhanced in the presence of genistein [22]. Cellular protein content was additively increased with 17β-estradiol and daidzein, but their effects on alkaline phosphatase activity were not additively. The receptors of estrogen are found in osteoblastic cells. Genistein has been shown to bind to estrogen receptor β in osteoblastic cells [18], and daidzein cannot bind to the receptor [18]. It is speculated that genistein may bind to estrogen receptor β in osteoblastic cells, although the nuclear localization of genistein in these cells has poorly understood.

DNA content in osteoblastic cells was increased in the presence of genistein or daizein, suggesting that isoflavones stimulate cell proliferation [22, 23]. Isoflavones increase alkaline phosphatase activity, which is a marker enzyme in the differentiation of osteoblastic cells. Genistein and daidzein may have the stimulatory effects on the proliferation and differentiation of osteoblastic MC3T3-E1 cells.

The cellular mechanism by which isoflavones stimulate osteoblastic bone formation has been studied in relation to protein synthesis [24]. The presence of genistein or daidzein stimulated protein synthesis in osteoblastic MC3T3-E1 cells *in vitro* [24]. The addition of genistein or daidzein into the reaction mixture of protein synthesis using the cell homogenate from osteoblastic cells cultured without isoflavone caused a significant increase in protein synthesis *in vitro* [24]. This increase was suppressed in the presence of an inhibitor of protein synthesis, indicating that genistein or daidzein can directly stimulate protein synthesis *in vitro*. Moreover, genistein or daidzein has been shown to increase [^{3}H]-leucyl-tRNA synthetase activity in the cytosol fraction of osteoblastic cell homogenate [24]. Genistein had a greater effect than daidzein [24]. Isoflavone can directly activate leucyl-tRNA synthetase, a rate-limiting enzyme in the translational process of protein synthesis.

The possibility that genistein or daidzein acts on the translational process in osteoblastic MC3T3-E1 cells cannot exclude, however, since the stimulating effects of the isoflavones on protein synthesis in osteoblastic cells are suppressed by actinomycin D, an inhibitor of transcription activity [24]. Genistein has been shown to bind to estrogen receptors in osteoblastic cells [13], although daidzein cannot bind to estrogen receptors [18]. Moreover, it is possible that genistein or daidzein can bind to other transcription factors, which differ from estrogen receptors, in osteoblastic MC3T3-E1 cells.

The estrogenic compounds of 17β-estradiol and genistein mediate very different actions on osteoblastic cells [25]. While 17β-estradiol may stimulate bone anabolism, in part, by antagonizing TNFα-induced NF-κB activation, genistein did not only fail to prevent the cytokin-induced NF-κB activation, but directly promotes NF-κB activation in MC3T3 cells [25]. These observations suggest an important mechanistic difference in the mechanisms by which 17β-estradiol and genistein promote osteoblast differentiation [25]. Whether genistein has an effect on various protein kinases and protein phosphatases that are related to osteoblastic cell proliferation and nuclear transcription activity remains to be elucidated.

Molecular Mechanism by Which Genistein Stimulates Osteoblastogenesis

The cellular and molecular mechanism by which genistein enhances osteoblastogenesis has been examined [26, 27]. Culture with genistein (10^{-5} M) has been found to have a stimulative effect on the mineralization in osteoblastic cells *in vitro* [26]. Alkaline phosphatase is an enzyme marker of osteoblasts and the enzyme participates in bone mineralization [27]. α1 (I) collagen is a matrix protein that is related to bone formation and mineralization in osteoblast lineage cells [28]. Osteocalcin is a bone matrix protein containing γ_2-carboxyglutamin acid, which is synthesized in osteoblasts and is protein involved in mineralization [28]. The expression of alkaline phosphatase, α1 (I) collagen, and osteocalcin mRNAs in osteoblastic cells was markedly enhanced after culture with genistein (10^{-6} or 10^{-5} M) as compared to the effect of each factor [26].

The effect of genistein (10^{-5} M) on the gene expression in osteoblastic cells was completely depressed in the presence of cycloheximide, an inhibitor of protein synthesis, and 5, 6-dichlorobenzimidazone-1-β-D-ribofuranoside (DRB), an inhibitor of transcriptional activity [26]. This suggests that genistein stimulates transcriptional process in osteoblastic cells. Genistein may have a stimulative effect on protein synthesis at the transcriptional process. Genistein (10^{-6} or 10^{-5} M) was found to increase protein content in osteoblastic MC3T3-E1 cells *in vitro* [26]. This effect of genistein may involve in the activation of aminoacyl-tRNA synthase, a rate-limiting enzyme in protein synthesis.

Genistein may stimulate both transcriptional and translational activities in osteoblastic cells. It is speculated that genistein has an effect on the binding of transcriptional factor(s), which is synthesized through translational process, to the nuclear DNA in osteoblastic cells. Genistein (10^{-6} M) has been shown to result in an increase in cell proliferation and stimulated osteoblastic maturation as assessed by culture duration-dependent increments in alkaline phosphatase activity, calcium deposition into extracellular matrix and Runx2/Cbfa1 gene expression in primary mouse bone marrow-derived mesenchymal stem cells (BMSCs) cultures [29]. Genistein also resulted in an increase in nitric oxide (NO) synthase activity, NO formation, and cyclic guanosine 3', 5'-monophosphate production in BMSCs cultures [29]. These effects of genistein were mimicked by 17β-estradiol [29]. Whether genistein and daidzein regulate target gene transcription through cyclic adenosine 3', 5'-monophosphate (cAMP) regulatory element (CRE) in human osteoblastic cell line MG-63 cells has been shown [30]. It was found that 17β-estradiol, genistein and daidzein suppressed the transcriptional activity of CRE-luciferase reporter gene in MG-63 cells [30]. 17β-estradiol and genistein but not daidzein inhibited the cAMP analogue 8-Br cAMP-induced transcription of CRE reporter gene [30]. Both genistein and 17β-estradiol inhibited basal and cAMP-induced mRNA levels of endogenous estrogen responsive genes containing CRE/CRE-like elements in their promoter regions, including interleukin (IL) 8 and serum- and glucocorticoid-inducible kinase 1 [30]. These findings indicate that genistein and daidzein may modulate bone remodeling through ERs by regulating target gene expression through the CRE motifs.

Genistein Suppresses Bone Resorption

Genistein Suppresses Bone Resorption in Bone Tissues

Parathyroid hormone (PTH), prostaglandin E_2 (PGE_2), and lipopolysaccharide (LPS) have a stimulative effect on bone resorption in *in vitro* culture system [31-33].

Culture with PTH, PGE_2, or LPS clearly stimulated bone resorption in the femoral-metaphyseal tissues cultured for 48 hours, when bone resorption was estimated with a decrease in bone calcium content [34]. The effects of bone-resorbing factors were completely suppressed in the presence of genistein, indicating an inhibition of bone resorption [34]. PTH and PGE_2 caused a remarkable increase in glucose consumption and lactic acid production by bone tissues [34]. The production of lactic acid by bone tissues may be partly related to the augmentation of glucose consumption.

Presumably, PTH- and PGE_2-stimulated lactic acid productions in bone tissues hasten a decrease in bone calcium content. Genistein completely depressed the PTH- or PGE_2-induced increase in both glucose consumption and lactic acid production in bone tissues [34]. The inhibitory effects of genistein on bone resorption may be partly related to the prevention of lactic acid production by bone tissues. Thus, genistein was demonstrated to possess suppressive effects on bone resorption in tissue culture *in vitro*. Moreover, genistein was found to reveal stimulatory effects on both bone formation and mineralization in tissue culture *in vitro*. Genistein may have an important role in the preservation of bone mass.

Genistein Suppresses Osteoclastogenesis

Osteoclasts, bone-resorbing cells, are formed from bone marrow stem cells [10, 11, 35]. Osteoclast-like cell formation is estimated using staining for tartrate-resistant acid phosphatase (TRACP), a marker enzyme of osteoclasts [35]. Mouse bone marrow cells were cultured for 7 days in a medium containing a bone-resorbing agent [PTH, PGE_2, 1, 25-dihydroxyvitamin D_3 (VD_3), or LPS] with an effective concentration [36]. Culture with PTH, PGE_2, VD_3, or LPS induced a remarkable increase in osteoclast-like multinucleated cells. These increases were depressed in the presence of genistein [36]. The suppressive effects of genistein (10^{-5} M) were equal to the effects of other anti-bone-resorbing agents (calcitonin, 17β-estradiol, and zinc sulfate) on osteoclast-like cell formation in mouse marrow culture [36].

Genistein has been shown to have a potent suppressive effect at the later stage of differentiation of bone marrow cells [36]. The suppressive effect of genistein on osteoclast-like MNC formation in mouse bone marrow culture was greater than that of daidzein.

The suppressive effect of genistein on osteoclast-like MNC formation induced by PTH or PGE_2 may be partly based on the blocking of the pathway of cyclic AMP signaling, but not that by protein kinase C signaling, at the differentiation stage of bone marrow cells [36]. In addition, if genistein can inhibit tyrosine kinase [15], the action of isoflavone on osteoclastic cell formation may be partly related to its inhibitory effect on the kinase. The cellular mechanism by which genistein inhibits osteoclast-like cell formation from marrow cells may also be involved in cyclic AMP signaling.

Interleukin-6 (IL-6) has been known to stimulate osteoclastogenesis and osteoprotegerin (OPG) depress binding of RANKL to RANK, which is receptor protein of RANKL [10]. These factors are produced in osteoblastic cells [10]. OPG mRNA expression has been found to increase in osteoblastic cells treated with 17β-estradiol, genistein, or daidzein [37]. The release of IL-6 in the cells was suppressed, while OPG production was enhanced by isoflavone treatments [37].

After adding the estrogen receptor-binding blocker, ICI-182780, the effects of isoflavones on OPG and IL-6 production disappeared [37]. Thus, the inhibition by dietary isoflavones of IL-6 production and the stimulation of OPG appear to be mediated, at least in part, via a genomic pathway operating through estrogen receptors and gene expression mechanisms [37]. The stimulation of OPG mRNA levels by genistein was not affected by the protein synthesis inhibitor, cycloheximide and was shown to be due to enhancement of OPG gene transcription [38]. Thus, genistein may partly suppress bone resorption and bone loss by enhanced osteoblastic production of OPG.

RANKL or tumor necrosis factor-α (TNF-α) has been shown to stimulate osteoclastic differentiation using RAW264.7, a murine monocytic cell line. The direct effects of estrogen on osteoclast precursor differentiation has been shown that, in addition to effecting osteoblasts, estrogen may protect bone by reducing osteoclast production [39].

Both genistein and estrogen has been found to inhibit RANKL-induced IkappaB degradation and NF-kappaB (NF-κB) nuclear localization, which is an intracellular signaling factor of RANKL, with the same relative potency as seen in proliferation and differentiation assays [39]. Genistein has been shown to depress TNF-α-induced c-fos and NFATc1 expressions in an estrogen receptor-dependent manner and also inhibit NFATc1 nuclear translocation [40]. This finding suggests that phytoestrogens suppress TNF-α-induced osteoclastogenesis via inhibition of c-fos-dependent NFATc1 expression [40].

Suppressive Effects of Genistein in Mature Osteoclasts

Genistein has been found to induce cell death (apoptosis) of osteoclasts isolated from rat femoral tissues [41, 42]. Calcitonin has been known to inhibit osteoclast activity [43]. This hormonal effect may be mediated through the two pathways of cyclic AMP and Ca^{2+} signalings [49]. The addition of dibucaine, an antagonist of calmodulin, or staurosporine, an inhibitor of pretin kinase C, completely prevented the decrease in osteoclasts induced by calcitonin, DcAMP, or calcium chloride [41]. The effect of genistein in decreasing osteoclasts was clearly suppressed by these inhibitors, suggesting that the effect of isoflavones on mature osteoclasts is partly mediated through the pathway of Ca^{2+} signaling.

A Ca^{2+} ionophore (A23187) caused a remarkable decrease in the number of osteoclasts, indicating that the entry of Ca^{2+} in the cells induces death of osteoclasts [44]. The number of osteoclasts was decreased, when were cultured in the presence of genistein or calcium chloride. Ca^{2+} can activate endonuclease and the metal induces apoptosis in cells [45]. The effect of A23187 in decreasing osteoclasts may be based on Ca^{2+}-activated DNA fragmentation, which induces apoptosis. Genistein has been shown to induce apoptosis in immature human thymocytes by inhibiting torosisomerase II [46].

Genistein may stimulate apoptosis of osteoclasts, and its mechanism is partly related to the pathway of Ca^{2+} signaling. It is possible that genistein stimulates Ca^{2+} entry into osteoclasts, since it has been reported that the isoflavone can directly open a chloride channel in human cystic fibrosis transmembranes [41].

In addition, daidzein has also been shown to suppress the number of mature osteoclasts [41], although daidzein did not have a greater suppressive effect than genistein. The effect of daidzein was abolished in the presence of dibucaine or staurosporine, supporting a partial involvement in the Ca^{2+}-signaling mechanism [41].

Culture with genistein has been found to cause an inhibition of protein kinase activity and an appreciable elevation of protein tyrosine phosphatase activity in osteoclasts [42]. Genistein has a strong inhibitory effect on protein tyrosine kinases [15]. The inhibition of these kinases may induce apoptosis in a human ovarian tumor cell line and in Jurket T-leukemia cells [15]. Genistein may partly induce apoptosis of osteoclasts through a mechanism that inhibits protein tyrosine kinases in the cells. It has been shown that tyrosine kinase Src is implicated in the process of osteoclast-induced bone resorption *in vitro* and *in vivo* [47].

Culture with genistein has also been found to cause a significant increase in protein tyrosine phosphatase activity in osteoclasts [42]. This effect was also seen after the addition of genistein into the enzyme reaction mixture *in vitro* [42]. Genistein can directly activate protein tyrosine phosphatase in osteoclasts. Vanadate is an inhibitor of protein tyrosine phosphatase [48]. Culture of osteoclasts in the presence of vanadate did not cause a decrease in the number of osteoclasts. The suppressive effect of genistein on osteoclasts was also seen in the presence of vanadate [42].

It has been known that protein tyrosine phosphatase (Src homology 2 domain-containing tyrosine phosphatase) is a negative regulator of osteoclastogenesis and osteoclast-resorbing activity in mutant mice [49]. The suppressive effect of genistein on mature osteoclasts may partly be mediated through the activation of protein tyrosine phosphatase in the cells.

β-Glucuronidase is a lysomal enzyme, which is involved in the stimulation of bone resorption by PTH. Genistein did not have an effect on β-glucuronidase activity in osteoclasts [42]. The suppressive effects of genistein on osteoclastic bone resorption may not be implicated in the activity of lysosomal enzymes in the cells.

Thus, suppressive effects of genistein on rat bone osteoclasts may be involved in cell apoptosis mediated through the Ca^{2+} siganaling mechanism, inhibition of protein kinase, and activation of protein tyrosine phosphatase in the cells.

Genistein Regulates Gene Expression in Osteoclasts

Genistein has been shown to regulate the gene expression of proteins that are related to osteoclast function. Culture with genistein (10^{-6} M) caused an increase in DNA fragmentation in isolated osteoclastic cells [50], indicating that genistein stimulates apoptotic cell death. In addition, the genistein-induced cell death was significantly prevented in the presence of caspase-3 inhibitor [50].

However, this effect was partial. Apoptotic cell death with genistein is partly mediated through activation of caspase-3, and also other mechanisms may participate in apoptotic cell death. The stimulatory effect of genistein (10^{-6} M) on the caspase-3 mRNA expression was completely suppressed after culture with cycloheximide, an inhibitor of protein synthesis, or DRB, an inhibitor of transcriptional activity, in osteoclastic cells with or without macrophage colony stimulating factor (M-CSF) and RANKL [50].

This finding suggests that the effect of genistein on caspase-3 mRNA expression is involved in protein synthesis and transcription activation in osteoclastic cells. The RANK activation by binding of RANKL on preosteoclasts stimulates the NF-κB signaling, propbably through association with TRAF-6 in osteoclasts [51, 52]. RANKL signaling targets include mitogen-activated protein kinases (MAPKs), NF-κB, and nuclear factor of activated T cells c1 (NFATc1) in osteoclast differentiation [51, 52].

Culture with genistein (10^{-6} M) did not cause a significant change in NF-κB mRNA expression in osteoclastic cells in the presence or absence of M-CSF and RANKL [44]. NFATc1 mRNA expression was significantly increased after culture with genistein (10^{-6} M) in osteoclastic cells cultured in the presence of M-CSF and RANKL [50]. Genistein may effect on the gene expression of NFATc1 that is a target protein of RANKL signaling in osteoclastic cells.

The decrease in NFATc1 expression with the combination of genistein may be important in the revelation of the suppressive effect on osteoclastogenesis and osteoclastic cell function.

The activation of TRACP or cathepsin K plays a role in the promotion of osteoclastic bone resorption by stimulating the decomposition of bone matrix [50].

Genistein (10^{-6} M) did not have a significant effect on TRACP mRNA expression in osteoclastic cells [50]. Culture with genistein was found to induce a significant decrease in cathepsin K mRNA expression in osteoclastic cells cultured in the presence or absence of M-CSF and RANKL [50].

This finding demonstrates that genistein (10^{-6} M) has a suppressive effect on the activity of cathepsin K in osteoclastic cells, which are activated through RANKL signaling. Genistein possesses a potent suppressive effect on enhanced bone-resorbing activity. As described above, genistein has been demonstrated to stimulate osteoblastic bone formation and inhibit osteoclastic bone resorption as summarized in Figure 1.

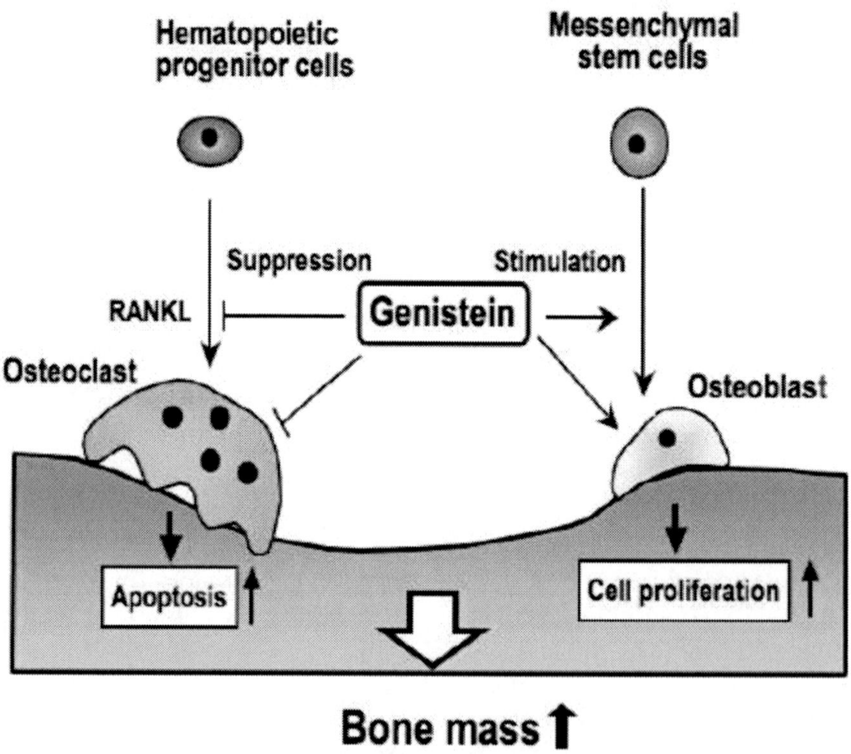

Figure 1. Regulatory role of genistein in bone homeostasis. Genistein stimulates osteoblastic bone formation and suppresses osteoclastic bone resorption, thereby increasing bone mass. Genistein stimulates protein synthesis and proliferation of osteoblastic cells, and it suppresses osteoclasogenesis from bone marrow stem cells and induces apoptosis of mature osteoclasts. Genistein regulates bone homeostasis through independent and dependent mechanisms on estrogen receptors in osteoblasts and osteoclasts. Genistein regulates the gene expression of various proteins that are related to osteoblastic bone formation and osteoclastic bone resorption. Genistein suppresses activity of various enzymes that stimulate bone resorption in osteoclasts. Moreover, intake of dietary soybean genistein prevents bone loss in overiectomized rats, which are an animal model for menopausal osteoporosis.

GENISTEIN PREVENTS OSTEOPOROSIS IN ANIMAL MODELS

Bone mass decreases with age. Osteoporosis with a decrease in bone mass is widely recognized as a major public health problem. This decrease may be due to increased osteoclastic bone resorption and to decreased osteoblastic bone formation, thereby increasing bone mass. Soybean genistein may prevent bone loss in osteoporosis.

Intake of dietary soybean genistein can reveal an anabolic effect on bone *in vivo* has been shown. *Nijiru,* which is a by-product in the fermentation process of soybean to make fermented soybean (*natto*), contains large quantities of natural isoflavone [53]. *Nijiru* may be a functional food factor as a dietary isoflavone.

Oral administration of the isoflavone-containing soybean extract (*nijiru*) to rats caused a significant increase in bone components (alkaline phosphatase activity, DNA and calcium contents) in the femoral-diaphyseal and -metaphyseal tissues *in vivo* [53], indicating the extract has an anabolic effect on bone metabolism. Also, *Nijiru* largely contains soybean saponin [54]. In addition, saponin has been shown to have a significant anabolic effect on bone components in the femoral tissues of rats [54].

Ovarian hormone deficiency at menopause stimulates bone loss. Ovariectomy (OVX) causes a lack of estrogen, and the hormone deficiency induces osteoporosis in humans [1-3]. Bone weight, bone mineral density, and bone mineral content were reduced in OVX rats [55].

Prolonged intake of fermented soybean diets with supplementation of isoflavone and saponin has been demonstrated to prevent bone loss in OVX rats [55], indicating that OVX-induced bone loss is prevented after the prolonged intake of dietary isoflavone and saponin supplementation.

Fermented soybeans are food sources that largely contain genistein. The effect of experimental diets using the fermented soybeans containing genistein on OVX-induced bone loss has also been demonstrated [56]. Experimental diets containing 44.6 to 92.4 mg of isoflavones (including genistein, genistein, daidzein, and daidzein) per 100 g of diet was fed to OVX rats for 3 months. OVX caused a significant reduction in the dry weight, mineral density, calcium content, and alkaline phosphatase activity in the femoral tissues [56]. These reductions were prevented after feeding a *natto* diet [56]. Thus, the prolonged intake of dietary *natto* supplemented with isoflavone has preventive effects on OVX-induced bone loss, suggesting that it may have a role in the prevention of osteoporosis.

The effect of a standardized soy extract (SSE, Soyselect) in the OVX rat model of menopause has also been examined [57]. OVX rats were orally treated for 6 weeks with the soy extract (50 or 100 mg/kg/day), vehicle (distilled water), or 17β-estradiol (0.5 mg/kg/day) [57]. The SSE produced a bone-sparing effect associated with a slowing down in the increased bone turnover observed after OVX (as indicated by measurements of serum osteocalcin levels and excretion ratio of deoxypyridinoline) [57]. Remarkably, organ weight data and histopathologic analysis did not show any stimulatory activity of the SSE on the uterus [57]. Immunohistochemical analysis showed a significant down-regulation of estrogen receptor-α in uterine epithelium after 17β-estradiol treatment, but not after treatment with the SSE [57]. After treatment with 17β-estradiol, estrogen receptor-β expression was not modulated in the stroma or epithelium, whereas the SSE induced an up-regulation of estrogen receptor-β stromal expression [57]. Collectively, these findings suggest that the lack of stimulatory activity on the uterine epithelium using soy treatment could be due to a negligible stimulatory activity on estrogen receptor-α and/or to the enhanced expression observed in stromal estrogen receptor-β, the latter being considered as a negative modulator of estrogen receptor-α-mediated uterine proliferation.

Uterine epithelial progesterone receptor was down-regulated by 17β-estradiol, but not the SSE, as compared with that in OVX rats [57]. The minor increase in lipid levels induced by OVX was not affected by SSE administration [57]. Thus, it has been demonstrated that the tested extract has an interesting profile of tissue-specific response, in that it is efficacious in preventing experimental osteoporosis without causing stimulation in uterus at doses that are effective in bone. As described above, intake of food factors including genistein and other isoflavones may be a useful tool in the prevention of osteoporosis with increasing ages.

Role of Genistein in Osteoporosis Prevention in Human Subjects

Preventive role of dietary genistein in osteoporosis has been demonstrated in human subjects. Twelve volunteers (six men and six women) were received *nijiru* twice a day for 60 days at a dose of 1500 mg (6 tablets) per day [58]. The serum concentration of γ-carboxylated osteocalcin, a marker of osteoblastic bone formation, was significantly increased to about 2-fold that in the control group after the intake of *nijiru* in both men and women [58].

The intake of isoflavone- and saponin-containing *nijir*u stimulated the γ-carboxylation of osteocalcin, which plays an important role in osteoblastic bone formation and mineralization, in healthy individuals [58]. The dietary intake of foods including genistein may have a role in the prevention of osteoporosis in human subjects.

Change in circulating biochemical markers of bone metabolism in aged individuals with the intake of fermented soybean (*natto*), which was made from isoflavone-rich soybean, has been examined [59]. Sixty-three volunteers (31 men and 32 women) were divided into four groups of 15 or 16 male volunteers and 16 or 16 female volunteers, and each group was sequentially given *natto* (40 g pack) containing once a day for 4 or 8 weeks as follows: either regular natto with naturally occuring isoflavone 35.0 mg and calcium 51.4 mg or supplemented *natto* containing isoflavone 35.0 mg and calcium 60.0 mg.

As serum bone markers, bone-specific alkaline phosphatase, γ-carboxylated osteocalcin, bone tartrate-resistant acid phosphatase (TRACP), and N-telopeptide of type I collagen were assayed. The intake of regular *natto* for 4 or 8 weeks in men or women persons caused a significant increase in γ-carboxylated osteocalcin, a marker of bone formation, and a significant decrease in serum bone N-teleopeptide of type I collagen, a marker of bone resorption, as compared with the value before intake [59]. Moreover, the intake of *natto* for 8 weeks in men or women caused a significant increase in serum bone-specific alkaline phosphatase activity and γ-carboxylated osteocalcin concentration and a significant decrease in serum bone TRACP activity and N-telopeptide of type I collagen, as compared with the values with the intake of regular *natt*o [59]. This study suggests that the intake of regular *natt*o with isoflavone-rich soybean has a stimulatory effect on bone formation and the suppressive effect on bone resorption in aged individuals.

Thus, prolonged intake of dietary genistein may be usefulness for bone health and the prevention of bone loss with aging and various pathophysiolocal states.

ANABOLIC EFFECTS OF GENISTEIN ON BONE ARE SYNERGISTICALLY ENHANCEDAND BY ZINC

Interestingly, the anabolic effects of genistein on bone metabolism have been shown to be synergistically enhanced by zinc, an essential trace element in growth [60]. Intake of combination with genistein and zinc is a useful tool in the prevention and treatment of osteoporosis with possessing of strong osteogenic effects. The recent findings with combination of genistein and zinc in bone metabolism are introduced in this section.

Role of Zinc in Bone Homeostasis

Zinc Plays a Pivotal Role As Zinc-Finger Protein in Bone Homeostasis

Zinc plays an important role in the regulation of bone homeostasis. Zinc finger transcription factors, which are zinc-binding protein, play a pivotal role in differentiation of osteoblastic cells and osteoclastic cells. Zinc transporter has been shown to locate in osteoblastic cells [61] and osteoclastic cells [62]. Zinc transporter may be a role in the uptake, intracellular sequestration or efflux of zinc. A novel zinc finger-containing transcription factor, called Osterix (Osx), has been found in osteoblastic cells [63]. Osx is required for osteoblast differentiation and bone formation and it acts downstream of Runx2/Cbfa1 [64]. Cas-interacting zinc finger protein (CIZ) has been found to be a novel type inhibitor of bone morphogenetic protein (BMP)/Smad signaling in the modulation of BMP2-induced osteoblastic cell differentiation [65]. Moreover, Schnurri-3 (Shn3) is an essential regulator of adult bone formation [66]. A novel TIZ (TRAF6-inhibitory zinc finger protein) has been shown to inhibit osteoclastogenesis and the function of tumor necrosis factor receptor-associated factor 6 [67]. Nutritional zinc status may influence on function of zinc finger proteins. Zinc is required for the growth, development, and maintenance of healthy bones. The retardation of bone growth is a common finding in various conditions associated with zinc deficiency [68, 69]. Skeleton contains a large proportion of the total body burden of zinc. Bone zinc has been shown to concentrate in the layer of osteoid prior to calcification [69]. Zinc deficiency is associated with many kinds of skeletal abnormalities in fetal and postnatal development. Nutritional zinc may play a physiologically important role in bone growth.

Moreover, zinc was found to be lower levels in the skeletal of osteoporotic patients than normal. In postmenopausal women, urinary zinc has been suggested as a marker of bone resorption, since women with osteoporosis excrete over than 800 μg zinc per g creatinine in urine [70].

Thus, zinc plays a pivotal role in the regulation of bone homeostasis.

Zinc Stimulates Osteoblastic Bone Formation

Zinc stimulates bone formation in *in vitro* and *in vivo* [71, 72]. Bone calcium content, alkaline phosphatase activity and collagen content have been shown to increase after culture with zinc, and these increases were depressed in the presence of an inhibitor of protein synthesis. Zinc has direct stimulatory effects on bone mineralization *in vitro*, and that bone protein synthesis is a necessary component of this response [73].

Endogenous zinc in the bone tissues has been shown to play an essential role in bone protein synthesis by using dipicolinate, a chelator of zinc [74]. Zinc has been shown to stimulate differentiation and proliferation in osteoblastic MC3T3-E1 cells [75, 76]. The stimulatory effect of zinc on protein synthesis in osteoblastic MC3T3-E1 cells [77] *in vitro* has been demonstrated using [^{3}H]-leucine.

The activity of [^{3}H]-leucyl-tRNA synthetase in the 105,000 *g* supernatant fraction (cytosol) of the bone homogenate was significantly increased about twofold after culture with zinc (10^{-4} M) [24, 28]. Zinc is found to activate [^{3}H]-leucyl-tRNA synthetase in the homogenate of osteoblastic cells [77]. Zinc has been shown to have direct stimulatory effects on protein synthesis at translational process. Endogenous zinc in the bone tissues is needed the stimulation of protein synthesis at the translational process in osteoblastic cells.

Zinc has been shown to increase various protein components including osteocalcin, IGF-I, or TGF-β1 in osteoblastic MC3T3-E1 cells [78]. The stimulatory effects of IGF-I on alkaline phosphatase activity, protein, DNA, and cell number in the cells were markedly enhanced in the presence of zinc sulfate (10^{-5} M) [79]. Such effect was not seen in case of both insulin and zinc. The enhancing effect of zinc on the effect of IGF-I may be mediated through signaling pathway of protein kinase C and protein phosphatase in osteoblastic cells [79]. Zinc increases protein tyrosine phosphatase activity in osteoblastic cells [80]. The effect of IGF-I in increasing this enzyme activity was enhanced after culture with zinc sulfate [80].

Zinc modulates anabolic effect of IGF-I on protein tyrosine phosphatase activity and cell proliferation. Zinc has also been to stimulate DNA synthesis in the homogenate of osteoblastic cells *in vitro* [81].

Moreover, zinc has been found to stimulate the mRNA expression of Runx2, a transcription factor, which is related to the differentiation from mesenchymal stem cells to preosteoblastic cells [82]. Thus, zinc stimulates cell differentiation, cell proliferation, and mineralization in osteoblasts, thereby promoting bone formation.

Zinc Suppresses Osteoclastic Bone Resorption

Zinc has a suppressive effect on bone resorption in tissue culture *in vitro* [83]. Calvaria were removed from weanling rats and cultured for periods of up to 48 hours in a medium containing various bone-resorbing factors [PTH, PGE_2, interleukin-1α (IL-1α), and LPS]. Culture with these factors caused a significant decrease in bone calcium content. Such decreases were completely inhibited in the presence of zinc (10^{-6} to 10^{-4} M). Also, zinc suppressed PTH- or IL-1α-induced increases in medium glucose consumpion and lactic acid production by bone tissues [83]. Thus, zinc has been demonstrated to have suppressive effects on bone resorption in bone tissue culture system *in vitro*.

Osteoclasts, bone-resorbing cells, are formed by differentiation of bone marrow cells.

Zinc has suppressive effects on osteoclast-like cell formation in mouse marrow culture *in vitro* [84, 85]. Presence of 1, 25-dihydroxyvitamin D_3 [1, $25(OH)_2D_3$], PTH, IL-1α, or PGE_2 induced a remarkable increase in osteoclast-like multinucleated cells [84]. These increases were suppressed in the presence of zinc [84]. Suppressive effects of zinc were equal in comparison with the effect of other anti-bone resorbing agents (calcitonin, 17β-estradiol, or acetazolamide) on osteoclast-like cell formation in mouse bone marrow culture [84].

In addition, culture with zinc caused apoptotic cell death of mature osteoclast-like cells isolated from rat femoral tissues [85]. Zinc has suppressive effects on osteoclastogenesis and osteoclastic cell death.

Stimulatory effects of PTH on osteoclast-like cell formation were weakened (about 50%) in the presence of Ca^{2+} chelator or dibucaine, a regulatory factor of intracellular Ca^{2+} signaling [86]. Phorbol 12-myristate 13-acetate (PMA), an activator of protein kinase C, stimulated osteoclast-like cell formation [86]. This effect of PMA was inhibited in the presence of zinc. These findings support the view that zinc inhibits PTH-stimulated osteoclast-like cell formation mediated through Ca^{2+}-dependent activation of protein kinase C. The receptor activator of NF-κB ligand (RANKL) plays a pivotal role in the development of osteoclasts from preosteoclasts [10]. RANKL is secreted from osteoblasts. RANKL is a member of the tumor necrosis factor (TNF) superfamily, which was originally identified as T-cell-derived immunomodulatory cytokines [10]. RANKL/RANK pathway is essential for osteoclast differentiation [10]. Expression of RANKL is increased in osteoblastic cells and bone marrow stromal cells in response to osteotropic factors such as PTH, 1, $25(OH)_2D_3$, or PGE_2. The effect of RANKL is abrogated by osteoprotegerin (OPG), a natural antagonist of RANKL, [10]. OPG is produced in osteoblastic cells. TNF receptor-associated factor (TRAF) family proteins are adaptor molecules. TRAFs bind to the membrane-proximal region of RANK and IL-1R-associated kinase and are critically involved in the intracellular signal transduction including NF-κB and mitogen-activated protein kinase (MAPK) activation [10].

Zinc has been shown to have suppressive effects on RANKL-induced osteoclast-like cell formation in mouse marrow culture in the presence of M-CSF [87]. Also, zinc inhibited TNF-α-induced osteoclastogenesis [87]. Suppressuve effects of zinc on osteoclastogenesis may be involved in inhibitory effect on RANKL stimulation. Culture with zinc has been shown to have stimulatory effects on the expression of OPG mRNA in osteoblastic cells [82]. The mechanism by which zinc suppresses osteoclastogenesis may also be related to production of OPG in osteoblastic cells.

As described above, zinc has been shown to have stimulatory effects on osteoblastic bone resorption and suppressive effects on osteoclastic bone resorption, thereby increasing bone mass. Zinc plays a pivotal in the regulation of osteoblasts and osteoclasts. This may be involved in many zinc-related proteins that are identified in bone cells.

Zinc Supplementation Prevents Bone Loss in Various Pathophysilogical States

Zinc has been shown to have preventive effects on bone loss in various pathophysiologic states. Fracture healing can be envisioned as involving five distinguishable processes, including the immediate response to injury, intramembranous bone formation, chondrogenesis, endochondral bone formation leading to the reestablishment of load bearing function, and bone remodeling [88].

These processes may occur simultaneously during fracture repair. During fracture healing, a number of growth factors, cytokines, and their cognate receptors are present at elevated levels in and around the fracture site [88]. The role of zinc in fracture healing has been examined using the diaphyseal tissues obtained at 7 or 14 days after the fracture of femoral diaphysis of rats [89, 90]. Oral administration of zinc acexamate (100 mg Zn/kg) for 28 days prevented the decrease in zinc and other bone components in the femoral-diaphyseal tissues of rats with fracture healing [91]. Zinc supplementation may have a role in the promotion of the healing of femoral fracture.

Zinc plays a role in the deterioration of bone metabolism with increasing age. Zinc content in the cellular components but not the matrix is lowered in the femoral diaphysis of elderly rats (30 weeks old) as compared with that of weanling rats (3 weeks old) [92].

Bone protein synthesis was most likely deteriorated with increasing age [92]. Oral administration of zinc sulfate (5 to 20 mg Zn/kg) for 3 days restored the rate of [^{3}H]-leucine incorporation [92]. Bone endogenous zinc may play a physiological role in the development of bone loss with increasing age. In aging, supplementation of zinc may be important in the prevention of bone loss with aging.

Zinc content is decreased in the femoral-metaphyseal tissues of rats with skeletal unloading, which is involved in the alteration of bone metabolism [93]. Skeletal unloading induces osteopenia after immobilization, spaceflight, bedrest, or hindlimb suspension. Skeletal unloading results in an inhibition of bone formation and induces an increase in bone resorption, thereby a loss of bone mass. Animals were fed for 4 days with the unloading. Unloading induced a decrease in metaphyseal zinc content [93]. Zinc accumulation in the metaphyseal tissues after a single oral administation of zinc sulfate (200 mg Zn/kg) was depressed with unloading [93].

β-Alanyl-histidinato zinc (AHZ) has been shown to have a potent effect on osteogenesis as compared with that of zinc sulfate [94]. Oral administration of AHZ has been shown to have a preventive effect on bone loss, which is caused with aging [95], inflammation [96], and ovariectomy [97].

Zinc acexamate has been found to have a potent effect on bone formation as compared with that of AHZ *in vitro*. Oral administration of zinc acexamate (25 mg Zn/kg body weight) for 14 or 21 days with once daily has been found to have restorative effects on the increase in serum glucose and triglyceride levels and the reduction of bone components induced in streptozotocin-induced diabetic rats [98, 99], indicating that zinc has a restorative effect on insulin-dependent diabetic conditions.

In addition, zinc yeast and citrated zinc bisglycinate has been developed as a zinc supplement. This is related to increase bioavailability in intestinal absorption of zinc, and it is used as a zinc supplementation. Supplemental intake with zinc, which has stimulatory effects on osteoblastic bone formation and mineralization and suppressive effects on osteoclastic bone resorption, is a useful tool in the prevention and treatment of osteoporosis.

Synergistic Effects of Genistein and Zinc in Osteoblastic Bone Formation

The effect of zinc sulfate (5.5 mg/kg body weight) on increasing DNA and calcium contents in the femoral tissues has been found to synergisctically enhance after simultaneous administration of genistein (100 μg/kg body weight) *in vivo* [100-102]. Oral administration of genistein (100 and 300 μg/kg) or zinc sulfate (1 and 5 mg Zn/kg) to female rats for 3 days caused a significant increase in alkaline phosphatase activity, DNA, and calcium content in the femoral-metaphyseal tissues *in vivo*. Combination of genistein and zinc may be a useful tool in the prevention and therapy of osteoporosis.

The cellular and molecular mechanisms by which zinc enhances anabolic effect of genistein on bone components have been examined. Combination of genistein (10^{-6} or 10^{-5} M) and zinc (10^{-5} or 10^{-4} M)) was found to have synergistic effects on protein content in osteoblastic MC3T3-E1 cells *in vitro* [103]. This effect may involve in the activation of aminoacyl-tRNA synthase by zinc and genistein, since these factors can increase the enzyme activity in the cytosol of osteoblastic cells *in vitro* [103]. Alkaline phosphatase is an enzyme marker of osteoblasts, and the enzyme participates in bone mineralization [27].

α1 (I) collagen is a matrix protein that is related to bone formation and mineralization in osteoblast lineage cells. Osteocalcin is a bone matrix protein containing γ_2-carboxyglutamin acid, which is synthesized in osteoblasts and is the protein involved in mineralization [28]. The expressions of alkaline phosphatase, α1 (I) collagen, and osteocalcin mRNAs in osteoblastic cells were synergistically enhanced after culture with the combination of genistein (10^{-6} or 10^{-5} M) and zinc (10^{-5} or 10^{-4} M) as compared to the effect of each factor [103]. This finding demonstrates that the combination of genistein and zinc has synergistic effects on gene expression in osteoblastic cells.

The effect of combination of genistein (10^{-5} M) and zinc (10^{-4} M) on the gene expression in osteoblastic cells was completely prevented in the presence of an inhibitor of protein synthesis or an inhibitor of transcriptional activity [103]. This suggests that the combination of genistein and zinc stimulates the transcriptional process in osteoblastic cells.

Genistein or zinc has stimulatory effects on protein synthesis at the transcriptional process. Presumably, combination of genistein and zinc stimulates both transcriptional and translational activities in osteoblastic cells. Each factor activates aminoacyl-tRNA synthetase at the transcriptional process in osteoblastic cells. Genistein or zinc may have an effect on the binding of transcriptional factor(s), which is synthesized through translational process, to the nuclear DNA in osteoblastic cells. Moreover, culture with genistein (10^{-5} M) or zinc (10^{-4} M) has stimulatory effects on mineralization in osteoblastic cells [103]. Combination of genistein and zinc synergistically enhanced mineralization in osteoblastic cells. This finding was supported the view that the combination of zinc and genistein can effectively enhance bone mineralization [103].

Synergistic Effects of Genistein and Zinc in Osteoclastic Bone Resorption

Moreover, combination of genistein (10^{-6} M) and zinc (10^{-5} M) has been shown to have potential suppressive effects on osteoclastic cells, which are generated in mouse marrow culture in the presence of M-CSF and RANKL *in vitro* [104]. The number of mature osteoclastic cells was synergistically decreased with culture of genistein (10^{-6} M) plus zinc (10^{-5} M) as compared with the value of each factor. Combination of genistein and zinc had potential suppressive effects on mature osteoclastic cells [104]. This effect was remarkable in the presence of M-CSF and RANKL, which are cytokines that stimulate osteoclastogenesis [104]. Culture with genistein or zinc caused a synergistical increase in DNA fragmentation in osteoclastic cells [104], indicating that both factors stimulate apoptotic cell death. In addition, genistein plus zinc-induced cell death was significantly prevented in the presence of caspase-3 inhibitor [104]. The stimulation of apoptotic cell death with the combination of genistein and zinc may be partly mediated through activation of caspase-3.

Zinc (10^{-5} M) increases the expression of caspase-3 mRNA in osteoclastic cells cultured with or without M-CSF and RANKL, while genistein (10^{-6} M) alone did not have an effect [104]. The effect of zinc was significantly enhanced with the combination of genistein.

Such an effect was completely inhibited after culture with an inhibitor of protein synthesis or an inhibitor of transcriptional activity in osteoclastic cells [104], suggesting that the effect with the combination of genistein and zinc is involved in protein synthesis and transcription activation in osteoclastic cells.

Activation of tartrate-resistant acid phosphatase (TRACP) or cathepsin K plays a role in the promotion of osteoclastic bone resorption by stimulating the decomposition of bone matrix [10, 11]. The expression of TRACP and cathepsin K mRNAs was markedly decreased in the presence of genistein plus zinc in osteoclastic cells cultured with M-CSF and RANKL [104]. Combination of genistein and zinc has potential suppressive effects on bone-resorbing activity. Thus, combination of genistein and zinc has been found to possess additive or synergistic effects in the suppression of osteoclastogenesis, mature osteoclast cell death, and bone resorption-related gene expression in osteoclastic cells.

Supplemental Intake of Genistein and Zinc Possesses Potential Effects against Bone Loss

As mentioned above, combination of genistein and zinc has been demonstrated to have potential synergistic effects on both stimulation of osteoblastic bone formation and suppression of osteoclastic bone resorption. Combination of genistein and zinc may have potential effects in the prevention and treatment of bone loss with various pathophysiologic states. Interestingly, genistein and zinc largely contain in fermented soybeans (*natto*). The effects of experimental diets with fermented soybeans containing genistein and zinc on OVX-induced bone loss has been demonstrated [56]. Experimental diets containing 2.1 to 9.7 mg of zinc per 100 g of diet and 44.6 to 92.4 mg of isoflavones (including genistein, genistein, daidzein, and daidzein) per 100 g of diet was fed to OVX rats for 3 months. OVX caused a significant reduction in the dry weight, mineral density, calcium content, zinc content, and alkaline phosphatase activity in the femoral tissues [56]. These reductions were prevented with feeding a *natto* diet. Such an effect was significantly enhanced in OVX rats fed a *natto* diet supplemented with zinc and isoflavone of more amounts. Prolonged intake of dietary *natto* supplemented with isoflavone and zinc has a preventive effect on OVX-induced bone loss, suggesting that it may have a role in the prevention and treatment of osteoporosis.

Change in circulating biochemical markers of bone metabolism in aged individuals with the intake of fermented soybean (*natto*), which was made from isoflavone-rich soybean, supplemented with zinc has been examined [59]. Sixty-three volunteers (31 men and 32 women) were divided into four groups of 15 or 16 male volunteers and 16 or 16 female volunteers, and each group was sequentially given *natto* (40 g pack) containing two different levels of zinc once a day for 4 or 8 weeks as follows: either regular *natto* with naturally occuring isoflavone 35.0 mg, zinc 0.8 mg and calcium 51.4 mg or supplemented *natto* containing isoflavone 35.0 mg, zinc 3.6 mg, and calcium 60.0 mg. As serum bone markers, bone-specific alkaline phosphatase, γ-carboxylated osteocalcin, bone tartrate-resistant acid phosphatase (TRACP), and N-telopeptide of type I collagen were assayed. The intake of regular *natto* for 4 or 8 weeks in men or women persons caused a significant increase in γ-carboxylated osteocalcin, a marker of bone formation, and a significant decrease in serum bone N-teleopeptide of type I collagen, a marker of bone resorption, as compared with the value before intake [59]. Moreover, the intake of zinc-supplemented *natto* for 8 weeks in men or women caused a significant increase in serum bone-specific alkaline phosphatase activity and γ-carboxylated osteocalcin concentration and a significant decrease in serum bone TRACP activity and N-telopeptide of type I collagen, as compared with the values with the intake of regular *natto* [59].

This study have demonstrated that the intake of regular *natt*o with genistein-rich soybean has stimulatory effects on bone formation and suppressive effects on bone resorption in aged individuals, and that this n effect is synergistically enhanced with supplementation of zinc. As described above, combination of genistein and zinc has been shown to possess synergistic effects on the stimulation of osteoblastic bone formation and the suppression of osteoclastic bone resorption. Supplementation with composition of pure genistein and zinc may be a potential tool in the prevention and therapy of osteoporosis with various pathophysiolocal conditions.

CONCLUSION

Bone homeostasis, which is regulated through osteoblastic cells and osteoclastic cells in bone tissues, is disordered with aging and pathophysiologic state including menopause.

This leads to bone loss, which may be due to decreased osteoblastic bone formation and increased osteoclastic bone resorption. Bone loss induces bone fracture, which is a public problem. Functional food factors may help to prevent bone loss with aging. This may be important in maintaining bone health in long life. Genistein has been demonstrated to stimulate osteoblastic bone formation and to suppress osteoclastic bone resorption. Fermented soybeans (*natto*) are good food sources that largely contain genistein. The dietary fermented soybean intake has a preventive effect in the prevention and treatment of bone loss in animal models and humans.

Moreover, the prevention of bone fracture with dietary fermented soybean intake is supported through epidemiologic investigation in Japan. Interestingly, soybean richly contains zinc, an essential trace element, which reveals a stimulatory effect on bone formation and a suppressive effect on bone resorption. Combination with genistein and zinc has been demonstrated to possess a synergistic-anabolic effect on bone mass. The composition of genistein and zinc has been developed as a novel supplement to prevent and treat osteoporosis and bone metabolic disorder. This biomedical food factor will be a useful tool in the prevention and treatment of osteoporosis in human subjects.

DISCLOSURES: The author has no conflicts of interest.

ACKNOWLEDGMENTS

The author was awarded from Mishima Kaiun Memorial Foundation, Japan, and Senji Miyata Foundation, Japan.

REFERENCES

[1] Cooper, C., Melton, J. III (1995) Epidemiology of osteoporosis. *Trends Endocrinol. Metab.* 3: 224-229.

[2] Riggs, B. L., Jowsey, J., Kelly, P. J., Jones, J. D., Maher, F. T. (1969) Effect of sex hormones on bone in primary osteoporosis. *J. Clin. Invest.* 48: 1065-1072.

[3] Weitzmann, M. N., Pacifici, R. (2006) Estrogen deficiency and bone loss: an inflammatory tale. *J. Clin. Invest.* 116: 1186-1194.

[4] Bonjour, J.-P., Schurch, M.-A., Rizzori, R. (1996) Nutritional aspects of hip fracture. *Bone* 18: 1395-1445.

[5] Yamaguchi, M. (2010) *Nutritional Factors and Osteoporosis Prevention*. Nova Science Publishers, Inc., New York, US.

[6] Yamaguchi, M. (2010) Role of nutritional zinc in the prevention of osteoporosis. *Mol. Cell. Biochem.* 338: 241-254.

[7] Yamaguchi, M. (2013) *Biomedical Osteoporosis Treatment*. Nova Science Publishers, Inc., New York, US.

[8] Parfitt, A. M. (1990) Bone-forming cells in clinical conditions. In: Hall, B. K., editor. *Bone* Volume 1. The Osteoblast and Osteocyte. Boca Raton, FL: Telford Press and CRC Press, pp. 351-429.

[9] Baron, R., Vignery, A., Horowitz, M. (1984) Lymphocytes, macrophages and the regulation of bone remodeling. *Bone Miner. Res.* 2:175-243.

[10] Zaidi, M., Blair, H. C., Moonga, B. S., Abe, E., Huang, C. L.-H. (2003) Osteoclastogenesis, bone resorption, and osteoblast-based therapeutics. *J. Bone Miner. Res.* 18: 599-609.

[11] Chambers, T. J., Fuller, K. (2011) How are osteoclasts induced to resorb bone? *Ann. N Y Acad. Sci.* 1240:1-6.

[12] Canalis, E., McCarthy, T., Centrella, M. (1988) Growth factors and the regulation of bone remodeling. *J. Clin. Invest.* 81: 277-281.

[13] Raggatt, L. J., Partridge, C. (2010) Cellular and molecular mechanisms of bone remodeling. *J. Biol. Chem.* 285:25103-25108.

[14] Chen, G., Deng, C., Li, Y. P. (2012) TGF-β and BMP signaling in osteoblast differentiation and bone formation. *Int. J. Biol. Sci.* 8:272-288.

[15] Liu, Y., Bhalla, K., Hill, C., Priest, D. G. (1994) Evidence for involvement of tyrosine phosphorylation in taxol-induced apoptosis in a human ovarian tumor cell line. *Biochem. Pharmacol.* 48: 1265-1272.

[16] Yamaguchi, M., Gao, Y. H. (1997) Anabolic effect of genistein on bone metabolism in the femoral-metaphyseal tissues of elderly rats is inhibited by the anti-estrogen tamoxifen. *Res. Exp. Med.* 197: 101-107.

[17] Yamaguchi, M., Gao, Y. H. (1998) Anabolic effect of genistein and genistein on bone metabolism in the femoral-metaphyseal tissues of elderly rats: The genistein effect is enhanced by zinc. *Mol. Cell. Biochem.* 178: 377-382.

[18] Kuiper, G. G., Lemmen, J. G., Carlsson, B., Corton, J. C., Safe, S. H., van der Saag, P. T., van der Burg, B., Gustafsson, J. A. (1998) Interaction of estrogenic chemicals and phytoestrogens with estrogen receptor beta. *Endocrinology* 139: 4252-4263.

[19] Gao, Y. H., Yamaguchi, M. (1999) Anabolic effect of daizein on cortical bone in tissue culture: Comparison with genistein effect. *Mol. Cell. Biochem.* 194: 93-98.

[20] Yamaguchi, M., Zhong, J. M., Fushimi, T. (2002) Anabolic effect of phosphogenistein and phosphodaidzein on bone components in rat femoral-metaphyseal tissues in vitro. *J. Bone Miner. Metab.* 20: 148-155.

[21] Yamaguchi, M., Ma, Z. J. (2001) Effect of polyphenols on calcium content and alkaline phosphatase activity in rat femoral tissues *in vitro*. *Biol. Pharm. Bull.* 24: 1437-1430.

[22] Sugimoto, E., Yamaguchi, M. (2000) Anabolic effect of genistein is osteoblastic MC3T3-E1 cells. *Int. J. Mol. Med.* 5: 515-520.

[23] Sugimoto, E., Yamaguchi, M. (2000) Stimulatory effect of daidzein in osteoblastic MC3T3-E1 cells. *Biochem. Pharmacol.* 59: 471-475.

[24] Yamaguchi, M., Sugimoto, E. (2000) Stimulatory effect of genistein and daizein on protein synthesis in osteoblastic MC3T3-E1 cells: Activation of aminoacyl-tRNA synthetase. *Mol. Cell. Biochem.* 214: 97-102.

[25] Yamaguchi, M., Weitzmann, M. N. (2009) The estrogen 17β-estradiol and phytoestrogen genistein mediate differential effects on osteoblastic NF-κB activity. *Int. J. Mol. Med.* 23: 297-301.

[26] Uchiyama, S., Yamaguchi, M. (2007) Genistein and zinc synergistically enhance gene expression and mineralization in osteoblastic MC3T3-E1 cells. *Int. J. Mol. Med.* 19: 213-220.

[27] Majeska, R. J., Wuthier, R. E. (1975) Studies on matrix vesicles Isolated from chick epiphyseal cartilage. Association of pyrophosphatase and ATPase activities with alkaline phosphatase. *Biochim. Biophys. Acta* 391: 51-60.

[28] Lian, J. B., Stein, G. S., Cannalis, E., Roky, P. G., Boskey, A. L. (1999) Bone formation: Osteoblast lineage cells, growth factor, matrix protein, and the mineralization process. In: Favus, M. J. (Ed.), *Primer on the Metabolic Bone Disease and Disorders of Mineral Metabolism* (4th edition), pp. 14-29, New York, Lippincott Williams and Wilkins Press.

[29] Pan, W., Quarles, L. D., Song, L. H., Yu, Y. H., Jiao, C., Tang, H. B., Jiang, C. H., Deng, H. W., Li, Y. J., Zhou, H. H., Xiao, Z. S. (2005) Genistein stimulates the osteoblastic differentiation via N0/cGMP in bone marrow culture. *J. Cell. Biochem.* 94: 307-316.

[30] Tang, X., Zhu, X., Liu, S., Wang, S., Ni, X. (2011) Isoflavones suppress cyclic adenosine 3',5'-monophosphate regulatory element-mediated trnscription in osteoblastic cell line. *J. Nutr. Biochem.* 22:865-873.

[31] Klein-Nulend, J., Fall, P. M., Raisz, L. G. (1990) Comparison of the effects of synthetic human parathyroid hormone (PTH)-(1-34)-related peptide of malignancy and bovine PTH-(1-34) on bone formation and resorption in organ culture. *Endocrinology* 126: 223-227.

[32] Graves, L. III, Jilka, R. L. (1990) Comparison of bone and parathyroid hormone as stimulators of osteoclast development and activity in calvarial cell cultures from normal and osteopetrotic (mi/mi) mice. *J. Cell. Physiol.* 145: 102-109.

[33] Klein, D. C., Raisz, L. G. (1970) Stimulation of bone resorption in tissue culture. *Endocrinology* 86: 1436-1440.

[34] Yamaguchi, M., Gao, Y. H. (1998) Inhibitory effect of genistein on bone resorption in tissue culture. *Biochem. Pharmacol.* 55:71-76.

[35] Takahashi, N., Yamana, H., Yoshiki, S., Roodman, G. D., Mundy, G. R., Jones, S. J., Boyde, A., Suda, T. (1988). Osteoclast-like cell formation and its regulation by osteotropic hormones in mouse bone marrow cultures. *Endocrinology* 122: 1373-1382.

[36] Gao, Y. H., Yamaguchi, M. (1999) Inhibitory effect of genistein on osteoclast-like cell formation in mouse marrow cultures. *Biochem. Pharmacol.* 58: 767-772.

[37] Chen, X. W., Garner, S. C., Anderson, J. J. (2002) Isoflavones regulate interleukin-6 and osteoprotegerin synthesis duering osteoblast cell differentiation via an estrogen-receptor-dependent pathway. *Biochem. Biophys. Res. Commun.* 295:417-422.
[38] Viereck, V., Grundker, C., Blaschke, S., Siggelkow, H., Emons, G., Hofbauer, L. C. (2002) Phytoestrogen genistein stimulates the production of osteoprotegerin in human trabecular osteoblasts. *J. Cell. Biochem.* 84:725-735.
[39] Garcia Palacios, V., Robinson, L. J., Borysenko, C. W., Lehmann, T., Kalla, S. E., Blair, H. C. (2005) Negative regulation of RANKL-induced osteoclastic differentiation in RAW264.7 cells by estrogen and phytoestrogens. *J. Biol. Chem.* 280:13720-13727.
[40] Karieb, S., Fox, S. W. (2011) Phtoestrogens directly inhibit TNF-α-induced bone resorption in RAW264.7 cells by suppressing c-fos-induced NFATc1 expression. *J. Cell. Biochem.* 112:476-487.
[41] Gao, Y. H., Yamaguchi, M. (1999) Suppressive effect of genistein on rat bone osteoclasts: Apoptosis is induced through Ca^{2+} signaling. *Biol. Pharm. Bull.* 22: 805-809.
[42] Gao, Y. H., Yamaguchi, M. (2000) Suppressive effect of genistein on rat bone osteoclasts: Involvement of protein kinase inhibition and protein tyrosine phosphatase activation. *Int. J. Mol. Med.* 5: 261-267.
[43] Murrills, R. J., Shane, E., Lindsay, R., Dempster, D. W. (1989) Bone resorption by isolated human osteoclasts in vitro: effects of calcitonin. *J. Bone Miner. Res.* 4: 259-268.
[44] Su, Y., Charkrabort, M., Nathanson, M. H., Baron, R. (1992) Differential effects of the 3',5'-cyclic adenosine monophosphate and proein kinase C pathways on the response of isolated rat osteoclasts to calcitonin. *Endocrinology* 131: 1497-1502.
[45] Jones, D. O. P., McConkey, D. J., Nicotera, P., Orrenius, S. (1989) Calcium-activated DNA fragmentation in rat liver nuclei. *J. Biol. Chem.* 264: 6398-6403.
[46] McCabe, M. J., Orrenius, S. (1993) Genistein induces apoptosis in immature human thymocytes by inhibiting topoisomerase-II. *Biochem. Biophys. Res. Commun.* 194: 944-950.
[47] Missbach, M., Jeschke, M., Feyen, J., Muller, K., Glatt, M., Green, J., Susa, M. (1999) A novel inhibitor of the tyrosine kinase Src suppresses phosphoryalation of its major cellular substrates and reduces bone resorption in vitro and in rodent models in vivo. *Bone* 24: 437-449.
[48] Hunter, T. (1995) Protein kinases and phosphatases: the Yin and Yang of protein phosphorylation and signaling. *Cell* 80: 225-236.
[49] Aoki, K., Didomenico, E., Sims, N. A., Mukhopadhyay, K., Neff, L., Houghton, A., Amling, M., Levy, J. B., Horne, W. C., Baron, R. (1999) The tyrosine phosphatase SHP-1 is a negative regulator of osteoclastogenesis and osteoclast resorbing activity: increased resorption and osteopenia in $me^{v}.me^{v}$ mutant mice. *Bone* 25: 261-267.
[50] Uchiyama, S., Yamaguchi, M. (2007) Genistein and zinc synergistically stimulate apoptotic cell death and suppress RANKL siganling-related gene expression in osteoclastic cells. *J. Cell. Biochem.* 101: 529-542.
[51] Asagiri, M., Takayanagi, H. (2007) The molecular understanding of osteoclast differentiation. *Bone* 40:251-264.

[52] Inoue, J., Ishida, T., Tsukamoto, N, Kobayashi, N., Naito, A., Azuma, S., Yamamoto, T. (2000). Tumor necrosis factor receptor-associated factor (TRAF) family: adaptor proteins that mediate cytokine signaling. *Exp. Cell. Res*. 254:14-24.

[53] Ono, R., Yamaguchi, M. (1999) Increase in bone components of rats orally administered isoflavone-containing soybean extract (Nijiru). *J. Health Sci.* 45: 66-69.

[54] Ono, R., Yamaguchi, M. (1999) Anabolic effect of soybean saponin on bone components in the femoral tissues of rats. *J. Health Sci.* 45: 251-255.

[55] Ono, R., Ma, Z. J., Yamaguchi, M. (2000) Prolonged intake of fermented soybean diets with supplementation of isoflavone and saponin prevents bone loss in ovariectomized rats. *J. Health Sci*. 46: 70-74.

[56] Ma, Z. J., Shimanuki, S., Igarashi, A., Kawasaki, Y., Yamaguchi, M. (2000) Preventiv effect of dietary fermented soybean on bone loss in ovariectomized rats: Enhancement with isoflavone and zinc supplementation. *J. Health Sci.* 46: 263-268.

[57] Gallo, D., Zannoni, G. F., Apollon, P., Martinelli, E., Ferlin, C., Passetti, G., Riva, A., Morazzoni, P., Bombardelli, E., Scambia, G. (2005) Characterization of the pharmacologic profile rat model of menopause: effects on bone, uterus, and lipid profile. *Menopause* 12:589-600.

[58] Yamaguchi, M., Ono, R., Ma, Z. J. (2001) Prolonged intake of isoflavone- and saponin-containing soybean extract *(nijiru)* supplement enhances circulating γ-carboxylated osteocalcin concentration in healthy individuals. *J. Health Sci.* 47: 579-58239.

[59] Yamaguchi, M., Igarashi, A., Sakai, M., Degawa, H., Ozawa, Y. (2005) Prolonged intake of dietary fermented isoflavone-rich soybean reinforced with zinc affects circulating bone biochemical markers in aged individuals. *J. Health Sci.* 51: 191-196.

[60] Yamaguchi, M. (2012) Nutritional factors and bone homeostasis: Synergistic effect of zinc and genistein osteogenesis. *Mol. Cell. Biochem.* 366:201-221.

[61] Inoue, K., Matsuda, K., Itoh, M., Kawaguchi, H., Tomoike, H., Aoyagi, T., Nagai, R., Hori, M., Nakamura, Y., Tanaka, T. (2002) Osteopenia and male-specific sudden cardiac death in mice lacking a zinc transporter gene, Znt5. *Hum. Mol. Genet.* 15:1775-1784.

[62] Khadeer, M. A., Sahu, S. N., Bai, G., Abdulla, S., Gupta, A. (2005) Expression of the zinc transporter ZIP1 in osteoclasts. *Bone* 37:296-304.

[63] Nakashima, K., Zhou, X., Kunkel, G., Zhang, Z., Deng, J. M., Behringer, R. R., de Crombrugghe, B. (2002) The novel zinc finger-containing transcription factor osterix is required for osteoblast differentiation and bone formation. *Cell* 108:17-29.

[64] Shen, Z. J., Nakamoto, T., Tsuji, K., Nifuji, A., Miyazono, K., Komori, T., Hirai, H., Noda, M. (2002) Negative regulation of bone morphogenetic protein/Smad signaling by Cas-interacting zinc finger protein in osteoblasts. *J. Biol. Chem.*, 277: 29840-29846.

[65] Jones, D. C., Wein, M. N., Oukka, M., Hofstaetter, J. G., Glimcher, M. J., Glimcher, L. H. (2006) Schnur Regulation of adult bone mass by the zinc finger adapter protein ri-3. *Science* 312: 1223-1227.

[66] Shin, J. N., Kim, I., Lee, J. S., Koh, G. Y., Lee, Z. H., Kim, H. H. (2002) A novel zinc finger protein that inhibits osteoclastogenesis and the function of tumor necrosis factor receptor-associated factor 6. *J. Biol. Chem.*, 277:8346-53.

[67] Hsieh, H. S., Navia, J. M. (1980) Zinc deficiency and bone formation in guinea pig alveolar implants. *J. Nutr.* 110:1581-1588.

[68] Oner, G., Bhaumick, B., Bala, R. M. (1984) Effect of zinc deficiency on serum somatomedin levels and skeletal growth in young rats. *Endocrinology* 114:1860-1863.
[69] Masters, D. G., Keen, C. L., Lonnerdal, B., Hurley, L. S. (1986) Release of zinc from maternal tissues during zinc deficiency or simultaneous zinc and calcium deficiency in the pregnant rat. *J. Nutr*. 116:2148-2154.
[70] Herzberg, M., Foldes, J., Steinberg, R., Menczel, J. (1990) Zinc excretion in osteoporotic women. *J. Bone Mineral*. Res. 5:251-257.
[71] Yamaguchi, M., Yamaguchi, R. (1986) Action of zinc on bone metabolism in rats. Increases in alkaline phosphatase activity and DNA content. *Biochem. Pharmacol*. 35: 773-777.
[72] Yamaguchi, M., Osishi, H., Suketa, Y. (1987) Stimulatory effect of zinc on bone formation in tissue culture. *Biochem. Pharmacol*. 36:4007-4012.
[73] Yamaguchi, M., Oishi, H., Suketa, Y. (1988) Zinc stimulation of bone protein synthesis in tissue culture. Activation of aminoacyl-tRNA synthetase. *Biochem. Pharmacol*. 37: 4075-4080.
[74] Yamaguchi, M., Matsui, R. (1989) Effect of dipicolinate, a chelator of zinc, on bone protein synthesis in tissue culture. The essential role of zinc. *Biochem. Pharmacol*. 38: 4485-4489.
[75] Hashizume, M., Yamaguchi, M. (1993) Stimulatory effect of β-alanyl-L-histidinato zinc on cell proliferation is dependent on protein synthesis in osteoblastic MC3T3-E1 cells. *Mol. Cell. Biochem*. 122:59-64.
[76] Hashizume, M., Yamaguchi, M. (1994) Effect of β-alanyl-L-histidinato zinc on differentiation of osteoblastic MC3T3-E1 cells: Increases in alkaline phosphatase activity and protein concentration. *Mol. Cell. Biochem*. 131:19-24.
[77] Yamaguchi, M., Kishi, S., Hashizume, M. (1994) Effect of zinc-chelating dipeptides on osteoblastic MC3T3-E1 cells: Activation of aminoacyl-tRNA synthetase. *Peptides* 15: 1367-1371.
[78] Yamaguchi, M., Hashizume, M. (1994) Effect of β-alanyl-L-histidinato zinc on proein components in osteoblastic MC3T3-E1 cells: Increases in osteocalcin, insulin-like growth factor-I and transforming growth factor-β. *Mol. Cell. Biochem*. 136:163-169.
[79] Matsui, T., Yamaguchi, M. (1995) Zinc modulation of insulin-like growth factor's effect in osteoblastic MC3T3-E1 cells. *Peptides* 16:1063-1068.
[80] Yamaguchi, M., Fukagawa, M. (2005) Role of zinc in regulation of protein tyrosine phosphatase activity in osteoblastic MC3T3-E1 cells: Zinc modulation of insulin-like growth factor-I's effect. *Calcif. Tissue Int*. 76:32-38.
[81] Yamaguchi, M., Matsui, T. (1996) Stimulatory effect of zinc-chelating dipeptide on deoxyribonucleic acid synthesis in osteoblastic MC3T3-E1 cells. *Peptide* 17:1207-1211.
[82] Yamaguchi, M., Goto, M., Uchiyama, S., Nakagawa, T. (2008) Effect of zinc on gene expression in osteoblastic MC3T3-E1 cells: enhancement of Runx2, OPG, and regucalcin mRNA expressions. *Mol. Cell. Biochem*. 312: 157-166.
[83] Yamaguchi, M., Segawa, Y., Shimokawa, N., Tsuzuike, N., Tagashira, E. (1992) Inhibitory effect of β-alanyl-L-histidinato zinc on bone resorption in tissue culture. *Pharmacology* 45:292-300.
[84] Kishi, S., Yamaguchi, M. (1994) Inhibitory effect of zinc compounds on osteoclast-like cell formation in mouse marrow cultures. *Biochem. Pharmacol*. 48:1225-1230.

[85] Yamaguchi, M., Kishi, S. (1996) Zinc compounds inhibit osteoclast-like cell formation at the earlier stage of rat marrow culture but not osteoclast function. *Mol. Cell. Biochem.* 158:171-177.

[86] Yamaguchi, M., Kishi, S. (1995) Inhibitory effect of zinc-chelating dipeptide on parathyroid hormone-stimulated osteoclast-like cell formation in mouse marrow cultures: Involvement of calcium signaling. *Peptides* 16:629-633.

[87] Yamaguchi, M., Uchiyma, S. (2004) Receptor activator of NF-κB ligand (RANKL)-stimulated osteoclastogenesis in mouse marrow culture is suppressed by zinc in vitro. *Int. J. Mol. Med.* 14:81-85.

[88] Eihorn, T. A. (1998) The cell and molecular biology of fracture healing. *Clin. Orthop.* 355S:S7-S21.

[89] Igarashi, A., Yamaguchi, M. (1999) Increase in bone protein components with healing rat fractures: Enhancement by zinc treatment. *Int. J. Mol. Med.* 4:615-620.

[90] Igarashi, A., Yamaguchi, M. (2001) Increase in bone growth factors with healing rat fractures: The enhancing effect of zinc. *Int. J. Mol. Med.* 8: 433-438.

[91] Igarashi, A., Yamaguchi, M. (1999) Stimulatory effect of zinc acexamate administration on fracture healing of the femoral-diaphyseal tissues in rats. *Gen. Pharmacol.* 32:463-469.

[92] Yamaguchi, M., Ozaki, K. (1990) Aging affects cellular zinc and protein synthesis in the femoral diaphysis of rats. *Res. Exp. Med.* 190:295-300.

[93] Yamaguchi, M., Ehara, Y. (1995) Zinc decrease and bone metabolism in the femoral-metaphyseal tissues of rats with skeletal unloading. *Calcif. Tissue Int.* 57:218-223.

[94] Yamaguchi, M., Ehara, Y. (1996) Effect of essential trace metal on bone metabolism in the femoral-metaphyseal tissues of rats with skeletal unloading: Comparison with zinc-chelating dipeptide. *Calcif. Tissue Int.* 59:27-32.

[95] Yamaguchi, M., Kishi, S. (1994) Effect of zinc-chelating dipeptide on bone metabolism in weanling rats: Comparison with β-alanyl-L-histidinato zinc-related compounds. *Peptides* 15:671-673.

[96] Kishi, S., Yamaguchi, M. (1994) Stimulatory effect of β-alanyl-L-histidinato zinc on alkaline phosphatase activity in bone tissue from elderly rats: Comparison with zinc sulfate action. *Biol. Pharm. Bull.* 17:345-347.

[97] Segawa, Y., Tsuzuike, N., Itokazu, Y., Tagashira, E., Yamaguch, M. (1993) Effect of β-alanyl-L-histidinato zinc on bone metabolism in rats with adjuvant arthritis. *Biol. Pharm. Bull.* 16:656-659.

[98] Kishi, S., Segawa, Y., Yamaguchi, M. (1994) Histomorphological confirmation of the preventive effect of β-alanyl-L-histidinato zinc on bone loss in ovariectomized rats. *Biol. Pharm. Bull.* 17:862-865.

[99] Yamaguchi, M., Uchiyama, S. (2003) Preventive effect of zinc acexamate administration in streptozotocin-diabetic rats: Restoration of bone loss. *Int. J. Mol. Med.* 12:755-761.

[100] Gao, Y. H., Yamaguchi, M. (1998) Zinc enhancement of genistein's anabolic effect on bone components in elderly female rats. *Gen. Pharmacol.* 31:199-202.

[101] Yamaguchi, M., Gao, Y. H., Ma, Z. J. (2000) Synergistic effect of genistein and zinc on bone components in the femoral-metaphyseal tissues of female rats. *J. Bone Miner. Metab.* 18:77-83.

[102] Ma, Z. J., Igarashi, A., Inagaki, M., Mitsugi, F., Yamaguchi, M. (2000) Supplemental intake of isoflavones and zinc-containing mineral mixture enhances bone components in the femoral tissues of rats with increasing age. *J. Health Sci.* 46:363-369.

[103] Uchiyama, S., Yamaguchi, M. (2007) Genistein and zinc synergistically enhance gene expression and mineralization in osteoblastic MC3T3-E1 cells. *Int. J. Mol. Med.* 19: 213-220.

[104] Uchiyama, S., Yamaguchi, M. (2007) Genistein and zinc synergistically stimulate apoptotic cell death and suppress RANKL signlaing-related gene expression in osteoclastic cells. *J. Cell. Biochem.* 101:529-542.

In: Genistein and Daidzein
Editor: Masayoshi Yamaguchi

ISBN: 978-1-63483-194-9

Chapter 4

CONTEMPORARY CLINICAL EFFECTS OF PHYTOESTROGENS ADMINISTRATION

***Milan Terzic*[1,2,*] *and Jelena Dotlic*[1,2]**

[1]Clinical Center of Serbia, Clinic of Obstetrics and Gynecology, Belgrade, Serbia

[2]Medical Faculty, University of Belgrade, Belgrade, Serbia

ABSTRACT

Phytoestrogens are natural polycyclic phenols found in certain plants that are structurally and functionally similar to 17ß–estradiol. When ingested and metabolized, they bind to estrogen receptors (mainly estradiol $ß_2$ receptors – ER $ß_2$) and produce estrogenic effects.

Furthermore, they also bind to peroxisome proliferator-activated receptor (PPAR) family and aryl hydrocarbon receptors. Moreover, phytoestrogens affect activity of tyrosine kinases, histones, transcription factors, AMP (Adenosine mono phosphate) pathways, RNA expression and other intracellular regulators.

Current research demonstrates that phytoestrogens are helpful in reducing the intensity and frequency of hot flushes in menopausal women. Still, complete elimination cannot be achieved, as phytoestrogens have weaker biological activity than endogenous and synthetic estrogens.

Certain phytoestrogens have also been shown to decrease vaginal atrophy. Moreover, unlike hormone therapy, phytoestrogens were not found to increase clotting risk.

Studies confirmed beneficial effects of phytoestrogens on endothelial cells, vascular smooth muscle and extracellular matrix cells. Due to nitrogen oxide (NO) production they decrease arterial stiffness and exert anti-atherosclerotic effects. Phytoestrogens also affect angiogenesis and reduce inflammation and tissue damage, which delays the progression of atherosclerosis.

* Corresponding author: Milan Terzic. Clinical Center of Serbia, Clinic of Obstetrics and Gynecology, Dr Koste Todorovica 26, Belgrade, Serbia. Medical Faculty, University of Belgrade, Dr Subotica 8, Belgrade, Serbia. E-mail: terzicmilan@yahoo.co.uk.

Furthermore, phytoestrogens increase high density lipoprotein (HDL) and decrease low density lipoprotein (LDL) concentrations in human plasma.

Neuroprotective effects of phytoestrogens are related to both their antioxidant properties and interaction with ER. Literature data show that phytoestrogens might improve cognition, prevent or at least postpone dementia.

By activating PPARγ receptors phytoestrogens down-regulate pro-inflammatory cytokines, such as Cyclooxygenase 2 (COX-2) and Inducible nitric oxide synthase (iNOS), increase reverse cholesterol transport and insulin sensitivity. Phytoestrogens increase energy expenditure by affecting AMP activated kinase signaling cascades, which inhibit adipogenesis. Therefore, they can exert different effects on metabolism and may complement conventional treatment for diseases linked to metabolic syndrome, polycystic ovary syndrome and diseases related to hyper-androgenism and obesity.

Phytoestrogens have shown antiresorptive effects on bone through ER signaling. Additionally, they also affect broad spectrum of molecular mechanisms supporting osteogenesis by inhibiting PPARγ, Sirtuin 1 and CCAAT-enhancer-binding proteins (C/EBP) α, the key adipogenic transcription factors. Stimulating main osteogenic transcription factors Core-binding factor subunit alpha-1 (CBF-alpha-1) and Osterix, they present themselves as a new strategy for treating bone loss.

Literature data strongly support the relationship between phytoestrogen diet supplementation and the lower risk for some cancers. Phytoestrogens have been shown to inhibit the carcinogenesis through the modulation of genes that control the cell-cycle progression. They inhibit the activation of the nuclear transcription factor, kappa light polypeptide gene enhancer in B-cells (NF-kappa B) and protein kinase B signaling pathway, which are known to maintain a homeostatic balance between cell survival and programmed cell death (apoptosis). Phytoestrogens are also implicated in reversion of epigenetic events observed in some cancers. Furthermore, phytoestrogens are known to have potent antioxidant properties (inhibition of oxidative stress) and are found to be potent inhibitors of angiogenesis and metastasis. Moreover, phytoestrogens regulate mucosal immune response by suppressing dendritic cell function.

In conclusion, phytoestrogens appear to exert numerous positive health effects and consequently are recommended for use in various clinical fields.

Keywords: Phytoestrogens, hot flushes, cardiovascular system, cognition, bone, metabolic syndrome, cancer

INTRODUCTION

Phytoestrogens are naturally occurring polycyclic phenols originating from different plants that are structurally and functionally similar to 17β–estradiol. Therefore, phytoestrogens can bind to estrogen receptors (ER) predominantly ERβ and activate transcription, which exerts similar, but somewhat weaker, effects than endogenous estrogens (Lorand et al. 2010, Molla et al. 2011).

Phytoestrogens are grouped into various classes such as coumestans, resorcylic acid lactones, flavonoids, stilbenes, mycotoxins, etc. However, the most important groups of phytoestrogens are isoflavones (formononetin, biochanin A, daidzein, genistein, O-desmethylangolensin-ODMA, equol, etc.) and lignans (enterodiol, enterolactone, matairesinol and secoisolariciresinol) (Terzic et al. 2012, Ko 2014).

Isoflavones are non-steroidal molecules with a 2-phenylnaphthaline-type chemical structure. Two isoflavones, daidzein and genistein, were found to have the most potent estrogenic activity, and, therefore, they are most widely used as food supplements and also the most thoroughly investigated (Sanchez-Borrego et al. 2014).

One of them, or both could be found in medications available worldwide (Menopace 40mg isoflavones and 0,7mg genistein: Vitabiotics, UK; Feminal 40mg isoflavones: Jadran Galenski Laboratorij, Croatia; Innoclim soybean extract with 40mg isoflavones: Innotech, France; etc.). Numerous plants are sources of isoflavones, such as legumes, red clovers, black cohosh, dong quai, kudzu root, lentils, kidney beans, lima beans, broad beans, Panax ginseng, licorice, wild Mexican yam, bran, cereal, chickpeas, flax seeds, whole wheat flour, bluegrass, some sorts of fruits and vegetables, tea and other cereal grains (Terzic et al. 2012).

The biggest supply of isoflavones nowadays is coming from soy which is ingested trough diet. The main soy isoflavones are daidzein and genistein (Kuhnle et al. 2009).

Content and bioavailability of isoflavones in human food can be decreased during extraction, processing and cooking. Non-fermented food has significantly more isoflavones than fermented food. Investigations have shown that baking of soy flour does not alter its isoflavone content (Ko 2014, Seidlova-Wuttke et al. 2013).

Pharmacokinetics and Metabolism of Isoflavones

Available literature data demonstrates that the absorption and metabolism of isoflavones is individual and variable (Terzic et al. 2012, Franke et al. 2014).

Most isoflavones are found in plants as glycoside-bound complexes and as such are ingested. In this form they are biologically inactive and need to undergo a series of metabolic changes within the body in order to become active. It seems that there are numerous metabolic pathways involved, but intestinal micro-flora contributes significantly to these processes (Sanchez-Borrego et al. 2014, Franke et al. 2014).

The beta-glycosidic forms are hydrolyzed to their aglycones in the intestine, where they are absorbed. One third of aglycon is absorbed as free isoflavone, and two thirds are converted to metabolites such as equol and p-ethylphenol by the intestinal micro-flora, which are also captivated (Zheng et al. 2013). Isoflavones undergo entero-hepatic circulation after absorption, and they are conjugated to glucuronic acid in the intestinal wall and liver. The glucuronidated metabolites of isoflavones are subjects that are found in blood (Yuan et al. 2007). Daidzein is further metabolized into two final substances: equol (good estrogenic activity) and ODMA (weaker estrogenic activity). The differences in intestinal micro-flora of people contribute to the different concentrations of metabolites produced, such influencing on estrogenic activity (Zheng et al. 2013). Individuals that produce equol (around 30% of people) have higher serum equol concentrations after daidzein ingestion. Genistein is metabolized to mostly estrogenically inactive p-ethylphenol (Yuan et al. 2007).

Each of the isoflavones appears to occur in various concentrations in different anatomic areas. Genistein, daidzein, and equol have been detected in urine, plasma, saliva, breast aspirate or cyst fluid and prostatic fluid (Jerome-Morais et al. 2011). Therefore, it is assumed that potential drug interactions are minimal, and significant dosage adjustments for renal or hepatic impairment are not necessary.

Nevertheless, some studies have shown limited absorption of isoflavones at higher doses. Consequently, further studies to establish the exact doses that should be given to treat different conditions are still necessary (Seidlova-Wuttke et al. 2013).

Currently, it is believed that isoflavones hormone-like activity occurs in the concentration range between 0.1-10 μM. Isoflavones should be taken in the daily amount of 100 mg, which is equivalent to 25 g of soy protein. The most commonly used daily doses are around 40 mg (Seidlova-Wuttke et al. 2013). The long-term use of high doses is not recommended, as consumption of 150 mg/day and more of isoflavones for 5 years may induce endometrial hyperplasia in some women (Messina 2010). Nevertheless, the safety of isoflavones was confirmed in numerous studies that have shown almost no side effects in women who were treated with suggested doses of isoflavones (Sanchez-Borrego et al. 2012).

Both free and metabolized isoflavones are excreted from the body around 24 hours after ingestion, predominantly by urine and, to a lesser extent, in feces (Pérez-Jiménez et al. 2010, Franke et al. 2014). Metabolism of isoflavones can also be altered by other dietary components. Carbohydrates rich diet causes extensive biotransformation of isoflavones with higher production of equol. Antibiotics that change intestinal micro-flora can reduce metabolism of isoflavones (Setchell et al. 2010).

Possible Mechanisms of Isoflavones Action

Isoflavones may exert effects through a variety of mechanisms. The major mode of action is through binding to estrogen receptors and mimicking estrogenic actions. Isoflavones are structurally similar to 17ß–estradiol and can be considered as natural selective estrogen receptor modulators (natural SERMs). They both have a pair of hydroxyl groups and a phenolic ring, which is required for binding to estrogen receptors (ER-α and ER-ß) (Seidlova-Wuttke et al. 2013, Ko 2014).

There are two types of estrogen receptors α and β. Both ERs are found in different tissues, but their expression varies. ERα is predominant in endometrium, breast cancer cells, ovarian stroma cells, hypothalamus and epithelium of the efferent ducts. On the other hand, ERβ is principally active in kidney, brain, bone, heart, lungs, intestinal mucosa, prostate, and endothelial cells. Studies have proven that the ratio of α- to β- ER subtype concentration can play a role in certain diseases (Ascenzi et al. 2006).

The estrogen receptors α and ß function as ligand-activated transcription factors. Once activated by joining with different ligants, they shift form cytoplasm into the nucleus. There they bind to estrogen response elements (ERE) in the promoter regions of target genes and initiate their transcription (Terzic et al. 2012).

The interactions between the two main ERs and their specific cofactors provide a basis for the tissue-selective actions of estrogens (Zhao et al. 2007). The actions of ER-α and ER-ß on gene transcription can be opposite. Studies have shown that ER-ß may exert estrogen action by directly modulating gene transcription or by modulating ER-α activity. ER-ß can function as a transcriptional inhibitor or activator depending on the agonist concentration (Terzic et al. 2012). Moreover, the different ER combinations may respond in another way to various ligands, which can translate into tissue selective agonistic and antagonistic effects (Shen et al. 2009).

After hydrolysis to the aglycone forms isoflavones bind to estrogen receptors, but with significantly less potency than endogenous estrogens. Moreover, isoflavones also bind with less potency to steroid-hormone serum-transport proteins (McCarty et al. 2006).

Isoflavones bind to both ERs but with a greater affinity for β subtypes (Bowers et al. 2000, Maggiolini et al. 2001). Moreover, genistein and daidzein were proven to induce the transcription of estrogen-responsive target genes to much greater extent when bound to ERß than when bound to ERα (Harris et al. 2005, Lehmann et al. 2005). Literature data suggest that equol has more potent estrogenic activity due to the fact that it binds with a greater affinity to ER than its precursor daidzein. However, genistein binds ERβ with greater affinity even than equol (Bovee et al. 2008, Setchell et al. 2010).

Like synthetic SERMs, isoflavones can act as both estrogen agonists and antagonists and their effects are both tissue related and individual (Jiang et al. 2010). Antiestrogenic properties can be partially explained by their competition with endogenous 17β–estradiol for ER (Setchell et al. 2001). ER mediated action of isoflavones can exert effect in two manners: classical and non-genomic. The first one involves initiating transcription of target genes through nuclear ERs, while the second non-genomic effects are mediated by membrane-associated or cytoplasmic ERs. The non-genomic mechanism of action leads to rapid cellular responses like nitric oxide release, calcium flux, and/or activation of different signaling pathways, such as the AMP-activated protein kinase (AMPK), mitogen-activated protein kinase (MAPK) and phosphoinositide 3- kinase (PI3K) (Sirotkin et al. 2003, Cederroth et al. 2012). Furthermore, isoflavones can interact with many enzymes as endogenous estrogens. They can even interfere with the process of estrogen metabolism (Moon et al. 2006).

Several isoflavones are known to modify the CYP450 enzyme system by either inducing or suppressing the transcription of CYP450 enzymes or by inhibiting or enhancing enzyme activity (Moon et al. 2006). Isoflavones can also affect steroid biosynthesis by inhibiting enzymes important for steroid biosynthesis, regulating aromatase and 5α reductase activity, inhibiting tyrosine kinase as well as regulating plasma sex hormone binding globulin (SHBG) levels (Cederroth et al. 2012, Ko 2014). Isoflavones may also act through pathways other than the interaction with estrogen receptors. These actions include inhibiting the transformation of estrone to estradiol, inhibiting pathways involved in cell growth as well as having antioxidant and numerous other activities (McCarty 2006). Moreover, they can bind to androgen, progesterone, oxytocin and peroxisome proliferator activator receptors. Isoflavones may also act as peroxisome proliferator-activated receptor gamma (PPARγ) ligands (Somponpun et al. 2002, Wang et al. 2010). Due to the fact that isoflavones bind to ER, thus provoking estrogen-like effects, they show impact on all tissues and organs where ERs are present. Moreover, as isoflavones interact better with the ERß than ERα, they can be used for reducing cardiovascular risk, osteoporosis prevention and treatment, reducing the menopausal symptoms and possible tumor prevention (Seidlova-Wuttke et al. 2013).

Isoflavone Mechanism of Action and Effects on Vasomotor Symptoms

Vasomotor symptoms that bother postmenopausal women are hot flashes and night sweats. Hot flushes represent a sudden sensation of heat or burning, which typically starts in the chest and rises to the neck and head. It is accompanied by profuse sweating, palpitations and redness of the skin caused by peripheral vasodilatation as a way of losing heat (Luoto 2010). The exact mechanism of hot flushes is complex and not well understood up to now. It is believed that flushes are caused by small temperature elevations within the dysfunctional thermoregulatory center and temperature regulating zone in the hypothalamus.

Out of all complex neuro-endocrine mechanisms involved in thermoregulation, estrogen withdrawal is thought to be the central factor in hot flushes start up (Vilar-Gonzalez et al. 2011). Therefore, hormone replacement is effective approach for hot flushes. However, adverse effects limit the use of these agents in various groups of individuals, such as patients with history of endometrial and breast cancer or venous thrombo-embolism. Therefore, different non-hormonal alternatives for the treatment of hot flushes have been assessed (Hill et al. 2010). Literature data regarding the management of vasomotor symptoms with isoflavones have been somewhat inconsistent, though overall promising. Systematic reviews and meta-analyses of placebo-controlled studies confirmed a significant reduction in flush frequency compared with placebo. Results of numerous trials found a reduction of daily hot flushes after 12 weeks of a genistein treatment. Treated women reported a decrease in both frequency and intensity of hot flushes (Chen et al. 2014).

On the other hand, some other studies did not find a significant improvement in vasomotor symptoms after use of isoflavones. Almost a half of patients still suffered from hot flushes after two years of isoflavones administration (Molla et al. 2011).

Finally, the third group of authors found that isoflavones appeared to significantly reduce the number of hot flushes, but to a lesser extent than hormone therapy (Roberts et al. 2014). Therefore, currently no final conclusions can be drawn regarding this matter. US National Institutes of Health (NIH) and The North American Menopause Society (NAMS) published a position statement that there is not sufficient data from clinical trials to either prove or disprove the efficacy of isoflavone supplements in the treatment of menopausal vasomotor symptoms. Nevertheless, it is reasonable to prescribe soy isoflavone supplements for menopausal vasomotor symptoms (Seidlova-Wuttke et al. 2013, Lagarib et al. 2014).

Finally, after taking into consideration all available literature data it can be said that soy isoflavones might in some extent reduce the frequency of hot flushes in women with four hot flushes per day or less. The minimal isoflavone dose providing significant benefits is 50 mg/day, but there is no linear dose response. Three months are needed to achieve the expected response to therapy, and if no improvement is registered within that time, therapy needs to be changed. If isoflavones are found to improve quality of life, therapy should be continued (Lagarib et al. 2014).

Isoflavone Mechanism of Action and Effects on Central Nervous System

Knowing that ER's are found throughout the brain including the regions important for learning, memory and higher-order cognitive function (hippocampus and prefrontal cortex) isoflavones might be used for prevention and/or treatment of age-related dementia (Soni et al. 2014).

Studies show that after one year of phytoestrogen intake, memory scores in investigated patients were slightly higher than those in the placebo group, but the finding is not significant. Scores on tests for verbal skills and tasks requiring concentration and visual attention were similar between the two groups (Carlson et al. 2008, Gleason et al. 2009). Literature data revealed positive correlation of isoflavone treatment with cognitive function (delayed recall memory) in perimenopausal women, but not significant or even negative associations were seen in women over 68 years of age. Therefore, authors hypothesized that isoflavones act on the "Window of Opportunity" base. Moreover, some studies imply on the dose-dependent effects of isoflavones on cognition (Soni et al. 2014). Possible neuro-protective effects of isoflavones are related to both their antioxidant properties and interaction with ER. However, more studies that would test the actual mechanism of action of isoflavones on cognition should be undertaken. There is also not enough evidence regarding dose and duration of isoflavone consumption (Sirotkin et al. 2014).

Postmenopausal women have a greater number of sleep disturbances (difficulty in initiating and maintaining sleep) as well as a lower sleep efficiency compared with women in reproductive period, which can have a negative impact on their quality of life. As sleep quality is linked to a variety of causes, the available literature data is limited by numerous influences and factors. However, positive effect of isoflavones on insomnia has been registered by some authors. Further research should be performed to confirm these findings (Bedell et al. 2014).

Isoflavone Mechanism of Action and Effects on Skin and Vaginal Atrophy

Estrogen significantly modulates skin physiology by binding to the receptors in keratinocytes, fibroblasts, melanocytes, hair follicles and sebaceous glands. Moreover, estrogen stimulates angiogenesis and immune responses in skin. On the other hand, estrogen levels decrease after menopause has adverse effect on collagen and elasticity of skin and mucosa, resulting in skin thinning, wrinkling, dryness and reduced vascularity, impaired wound healing, hair loss, pigmentary changes and increased incidence of skin malignancies (Bedell et al. 2014).

Isoflavones have anti-aging effect on the skin through ER binding, thus leading to better skin vascularization, cell proliferation, decreased oxidative stress and apoptosis. Additionally, isoflavones can directly promote production of hyaluronic acid, collagen and extracellular matrix proteins (Sirotkin et al. 2014). Vulvo-vaginal atrophy is quite common after menopause, with prevalence ranging from 10 to 50% of women. It causes vaginal dryness and decreased lubrication, irritation, itching, dyspareunia and urinary incontinence that can all deteriorate quality of life of menopausal women and negatively affect their sexual activities. While trials of HRT continue to show a positive effect on vaginal atrophy, studies on isoflavones are confronting (Bolego et al. 2003). Some studies found an increase in vaginal cell maturation upon treatment with isoflavones. Conversely, other authors observed no significant difference maturation and karyo-pyknotic index of the vaginal epithelium as well as vaginal pH. However, the symptoms of stress-incontinence and vaginal dryness were less severe in women on therapy compared to those who did not receive isoflavones. But, the overall mild protective effect is observed in treated patients. Current standpoint is that a longer period (at least three to four months) of isoflavone intake is necessary to establish positive effects on vaginal epithelium and its dryness (Bedell et al. 2014).

Table 1. Activities of isoflavones genistein and daidzein on skin and mucosa

Target structure	Outcome
keratinocytes, fibroblasts, melanocytes, hair follicles and sebaceous glands	production of hyaluronic acid, collagen and extracellular matrix proteins
ER	stimulation of angiogenesis, immune responses and cell proliferation
	decreased oxidative stress and apoptosis

ISOFLAVONE MECHANISM OF ACTION AND EFFECTS ON BONES

Bone remodeling is mediated by osteoclasts (responsible for bone resorption) and osteoblasts (responsible for bone formation).

The function of both osteoblasts and osteoclasts can be altered by different cytokines, growth factors and many hormones out of which estrogen has the main role in women (Lagarib et al. 2014). Estrogen has multiple other effects that relate to the skeleton like enhanced intestinal calcium absorption or protection of bones from the resorptive effects of PTH. In menopause with estrogen depletion, osteoclast activity exceeds osteoblast activity, leading to a rapid loss of bone mineral content as well as matrix, and an increased risk of fractures (Krum et al. 2008).

When current data was assessed in undertaken meta-analyses, the benefits of isoflavones on bone mass were confirmed. A literature review revealed that majority of authors found a positive correlation between isoflavones and bone health of both peri- and postmenopausal women starting from 6 months after isoflavone treatment commencement. However, it was suggested that benefits may be specific to skeletal areas. Isoflavones can especially improve lumbar spine bone mineral density (BMD) (Bedell et al. 2014, Chen et al. 2014). Therefore, the North American Menopause Society (NAMS) published a report stating that it is reasonable to prescribe isoflavones for osteoporosis prevention and/or treatment in peri- and postmenopausal women. Daily dose of 90 mg for at least six months is needed for effective treatment. If there is improvement treatment should be continued, but if the symptoms do not resolve after 12 weeks other alternatives need to be considered (Lagarib et al. 2014).

The mechanisms by which isoflavones influence bone remodeling are still under discussion, but it is considered that isoflavones can act by binding to estrogen receptors as well as via nonestrogenic mechanisms. Evidence from recent studies suggests that isoflavones can act in multiple ways on both osteoblasts and osteoclasts in order to maintain bone mass (Dai et al. 2013).

Antiresorptive effects of isoflavones on bone can be mediated through ER signaling similarly to endogenous estrogen (activation of genes for osteoclast apoptosis). Furthermore, in women with osteoporosis proinflammatory cytokine – interleukin 6 (IL-6) is increased. Isoflavones can inhibit transcription factor nuclear factor κB (NF-κB) activation that consequently leads to reduction in IL-6 production (Dijsselbloem et al. 2004). Additionally, they also affect broad spectrum of molecular mechanisms supporting osteogenesis by inhibiting PPARγ, Sirtuin 1 and CCAAT-enhancer-binding proteins (C/EBP) α, the key adipogenic transcription factors. Stimulating osteoblastic differentiation and main osteogenic transcription factors Core-binding factor subunit alpha-1 (CBF-alpha-1) and Osterix, they present themselves as a new strategy for treating bone loss (Liao et al. 2007).

Bone morphogenetic proteins (BMPs) are members of the TGF-β superfamily that seem to be key regulators of bone formation and remodeling. In some in vitro studies it was observed that BMP 1 to 8 transcripts were up-regulated shortly after initiating genistein treatment. It was reported that the expression of BMP2 and BMP6 in promoting osteoblast differentiation and mineralization was significantly increased via ERα-dependent mechanism. Moreover, fibroblast growth factors (FGFs) and colony stimulating factors (CSFs) were also increased after genistein treatment (Liao et al. 2007, Dai et al. 2013). Osteoclast activity is regulated by phosphorylation of cell membrane constituents, involving tyrosine kinases. Genistein is one of the naturally occurring tyrosine kinase inhibitors and therefore can impede osteoclastic bone resorption. Isoflavones act on osteoclasts also by inhibition of topoisomerase I and II and induction of their apoptosis (Bitto et al. 2010).

Table 2. Activities of isoflavones genistein and daidzein on bones

Target structure/activity	Outcome
ER signaling	activation of genes for osteoclast apoptosis
inhibition of transcription factor nuclear factor κB (NF-κB)	reduction of IL-6 production with consequent anti-inflammatory effects
inhibiting adipogenic transcription factors PPARγ, sirtuin 1 and CCAAT-enhancer-binding proteins (C/EBP) α	osteogenesis
stimulation core-binding factor subunit alpha-1 (CBF-alpha-1) and Osterix	
production of osteoprotogerin (OPG) and neutralization of RANKL	osteoclasts deactivation
increasing DNA and protein content and alkaline phosphatase activity	osteoblasts stimulation

Daidzein can have stimulatory effect on osteoblasts by increasing DNA and protein content as well as alkaline phosphatase activity. Osteoblasts express the receptor activator of nuclear factor-κB ligand (RANKL). RANKL is involved in osteoclast differentiation in a paracrine manner by binding to its receptor, which is found on the osteoclast membrane. Osteoblasts also secrete osteoprotogerin (OPG), which can neutralize the actions of RANKL. Estrogen increases the production of OPG and hence decreases the activity of RANKL. Genistein can also increase the production of OPG from osteoblasts, providing additional bone-sparing (Marini et al. 2008).

Isoflavone Mechanism of Action and Effects on Cardiovascular System

It is well known that the risk for cardiovascular disease in women increases after menopause mostly due to the loss of above described protective antioxidant estrogen function

(Arnal et al. 2007). The classic risk factors of atherosclerosis, such as hypertension, hypercholesterolemia and smoking, increase the oxidative stress in the arterial wall. The response of endothelial cells is activation and inflammation, with increased level of C-reactive protein (CRP) and expression of cellular adhesion molecules (CAM).

Estrogen was found to reduce the concentrations of CAM as well as other markers of oxidative stress (De Andrade et al. 2014). Isoflavones, being of similar structure and activity as estrogen, might also be used for cardio protection. An increasing number of studies proved the beneficial effect of soy on metabolic syndrome. Numerous mechanisms are included: down regulation of pro-inflammatory cytokines COX-2 and iNOS through PPAR activation and IκB inhibition, PPARγ mediated augmentation of reverse cholesterol transport, PPARα mediated increase of insulin sensitivity, activation of antioxidant genes through KEAP, eNOS activation, inhibition of adipogenesis by AMP-activated kinase signaling cascades that increase energy outflow, etc (Bolego et al. 2003, Sirotkin et al. 2014).

Literature data show that genistein and daidzein can be involved in prevention of endothelial lesion development and coronary artery atherosclerosis. Isoflavones might induce vascular remodeling and neointimal development in damaged endothelium by inhibiting mitogen-induced proliferation, migration and extracellular matrix synthesis and down-regulated MAP kinase activity (De Andrade et al. 2014).

Additionally, genistein and daidzein were found to act as anti-thrombotic and anti-atherogenic agents as they decrease monocyte chemo-attractant protein-1 and collagen-induced platelet aggregation. Moreover, they induce vasodilatation with similar potency to 17ß-estradiol by stimulating the activity of the endothelial NO synthase (NOS) as well as enhancing reaction to acetylcholine (Gottstein et al. 2003, Liu et al. 2004).

Isoflavones also exert a direct effect on cardiac myocyte contractility. Genistein and daidzein were found to cause significant inhibition of calcium flow in myocytes (similarly to effects of calcium channel blockers) and relax myocytes (Bolego et al. 2003).

Significant changes in serum and plasma lipoprotein levels (high density lipoprotein-HDL, low density lipoprotein-LDL, tryglicerides, cholesterol) that occur in menopausal women are clear predictors of cardiovascular risk, morbidity and mortality (Polotsky et al. 2010). Based on all available literature, treatment with isoflavones was found to provoke lipid-lowering effects with a significant decrease in total cholesterol, LDL cholesterol and triglycerides (Torp et al. 2008, Terzic et al. 2009).

The results regarding HDL cholesterol are inconsistent. While some meta-analyses of interventional studies observed no significant change in HDL-cholesterol, others found small but yet significant increase in serum HDL cholesterol. However, even the sustained increase in HDL cholesterol is encouraging improvement in the lipid profile (Sirtori et al. 2007, Terzic et al. 2012).

Moreover, it should be pointed out that the benefits of isoflavones on lipid profiles are influenced by the baseline cholesterol concentration and the effects are even more pronounced in the setting of hypercholesterolemia (Bedell et al. 2014).

The mechanisms responsible for the effects of soy isoflavones on the lipid profile are still being explored. Isoflavones contained in intact soy protein may serve as a natural selective estrogen receptor modulator and exert an effect on lipid metabolism through their biological similarities to estrogens and estrogen-receptor-dependent gene expressions (Vogelvang et al. 2006).

Authors have proved the increase of LDL receptor activity in monocytes from hypercholesterolemic individuals after treatment with genistein and daidzein. Other possible mechanisms include effects on hepatic lipase activity, adipose tissue, amino acid profile and regulation of cholesterol homeostasis in Hep G2 cells, although more research is needed regarding these reactions (Kim et al. 2011).

Inconsistent results were obtained from several randomized controlled trials regarding the role of isoflavones on glucose metabolism as cardiovascular risk factors (Jungbauer et al. 2014). Literature data show that isoflavone use did not significantly lower glucose blood levels in perimenopausal and postmenopausal women.

On the other hand, the significant improvement of insulin secretion and resistance as well as overall metabolic homeostasis was found by some authors. Genistein was found to increase glucose sensitivity of pancreatic islets in a dose-dependent manner and independently of ER (Choi et al. 2009). Therefore, isoflavones might have overall beneficial effects, but further research should also be performed for definitive conclusions on this matter.

Table 3. Activities of isoflavones genistein and daidzein on cardiovascular system

Mechanism of action	Outcome
down regulation of pro-inflammatory cytokines COX-2 and iNOS	anti-atherosclerotic effects
activation of PPARα and PPARγ	increase of insulin sensitivity and augmentation of reverse cholesterol transport
down-regulation of MAP kinase	vascular remodeling and neointimal development
decrease of monocyte chemo-attractant protein-1 and collagen-induced platelet aggregation	anti-thrombotic and anti-atherogenic activity
stimulation of endothelial NO synthase (NOS) and enhancing reaction to acetylcholine	vasodilatation
KEAP activation	antioxidant effects
AMP-activated kinase signaling cascades	inhibition of adipogenesis
blocking of calcium channels	myocytes relaxation
enhancement of LDL receptor	lipid-lowering effects
hepatic lipase, adipose tissue and Hep G2 cells	regulation of cholesterol homeostasis

Isoflavone Mechanism of Action and Effects on Reproduction

According to results of the studies that tested effects of isoflavones on estrogen biosynthesis and estrogen biosynthetic enzymes it can be concluded that isoflavones influence on fertility. Decreased plasma concentrations of follicle-stimulating hormone (FSH), luteinizing hormone (LH) and progesterone as well as decreased serum concentrations of E2 and estrone were reported after isoflavone intake. However, menstrual cycle length and regularity was not changed (Kurzer 2002, Pérez-Jiménez et al. 2010).

Some authors found that isoflavone treatment may decrease FSH and LH levels in premenopausal women and increase estradiol in postmenopausal women. The clinical implications of these hormonal changes remain to be further determined (Cederroth et al. 2012).

Combined treatment with genistein and daidzein has adverse effects on human sperm and can lead to decreased sperm production. Genistein causes increased cAMP stimulation, leading to acrosomal loss and an increased rate of capacitation, resulting in inability of the sperm to fertilize an ovum (Fraser et al. 2006).

On experimental model, potential unfavorable effects on fertility have been observed in animals that were feed with soy and red clover. Studies in pregnant animals have shown that high doses of phytoestrogens can change the fetal hormonal environment (Cornwell et al. 2004). On the other hand, no adverse reproductive or endocrine effects were noted in children who were fed by soy-based formula as infants (Strom et al. 2001).

Moreover, soy infant formula with isoflavones was not found to affect the onset either puberty or sexual maturation. Current standpoint is that available data are inappropriate to estimate if soy formula has developmental or reproductive toxicity, but the overall risk seems to be minimal (Rozman et al. 2006).

Polycystic ovary syndrome (PCOS) is characterized by oligo-anovulation, hyper-androgenism and insulin sensitivity deterioration. PCOS seems to be a genetic disease that is linked to insulin and PPARγ receptor polymorphisms. Genistein and daidzein are weak peroxisome proliferator-activated receptor (PPAR) activators. Therefore, it was hypothesized that isoflavones may be helpful in treating PCOS. Further clinical studies are needed to examine whether these promising compounds are effective for treating this relatively frequent disorder (Cederroth et al. 2012).

Isoflavone Mechanism of Action and Effects on Cellular Growth and Proliferation

Concerns were raised that isoflavones could promote cell growth and in such manner stimulate growth of hormone related pre-existing tumors through activation of estrogen receptors. However, contrary to what might have been expected, isoflavones were found to have numerous antiproliferative properties. This could be explained by the distinctive activities of the ER isoforms as well as the specific affinities of isoflavones for ER-ß over ER-α. Moreover, isoflavones can alter the expression of a number of proteins that control cell cycle and induce cell cycle arrest and apoptosis (McCarty 2006).

Isoflavones might have a biphasic effect on cell proliferation that is dose-dependent.

Genistein was found to increase growth in estrogen-sensitive cells at low concentrations and decrease cell growth by suppressing DNA synthesis and inducing apoptosis at high concentrations (Maggiolini et al. 2001). Additionally, although isoflavones promote cell growth in ER-positive cells, ER-negative cells may have different responses to isoflavone exposure. Experimental studies found that genistein reduced proliferation or had no stimulatory effect on ER-negative breast carcinoma cell lines (Van der Woude et al. 2005).

Isoflavone Mechanism of Action and Effects on Malignant Tumors

Dietary chemopreventive agents have received significant attention in the area of cancer research (Lepri et al. 2014). It is believed that development of at least some cancers could be avoided by healthy diet, participating in regular physical activity and maintaining adequate body mass index (BMI). Characteristics of isoflavones genistein and daidzein suggest their ability to reduce cancer risk. Moreover, isoflavones have shown neither side effects nor toxicity even after prolonged use, and therefore their administration as chemopreventive substances can be considered safe (Lepri et al. 2014).

Several mechanisms have been hypothesized to explain the potential protective effects of isoflavones on cancers, but they can all be divided into two major groups: ER dependent and ER independent.

a) ER Dependent Pathway

By binding to estrogen receptors (ER) isoflavones can modify the sex hormone metabolism in hormone-related tumors (Neill et al. 2014). Isoflavones genistein and daidzein down-regulate ERα expression and inhibit the activity of enzymes involved in estrogen metabolism, such as aromatase and 17β-oxidoreductase (Staar et al. 2005). They can also increase plasma sex hormone binding globulin levels (SHBG) and in that manner reduce plasma levels of free biologically active sex hormones, which can diminish hormone-related tumors proliferation. In tissues with ER, genistein and daidzein have been shown to change the expression of ERα and ERβ-mRNA (Staar et al. 2005). Binding of genistein and daidzein to ER triggers signals inside cells that modify the ERα/β-mRNA expression in favor of higher response on ERβ. Moreover, phytoestrogens might enhance the apoptotic effect by stimulating ERα (Sakamoto et al. 2010).

b) ER Independent Pathway

Anticancerous effects could also be mediated through ER-independent pathways (Staar et al. 2005). The possible biological mechanisms by which flavonoids exert their anticarcinogenesis effects rely on antiproliferative activity and cell-cycle arrest, induction of apoptosis, modulation of phase I and II metabolizing enzymes, inhibition of angiogenesis and metastasis and antioxidative and anti-inflammatory properties (Lepri et al. 2014).

Antiproliferative Activity, Cell-Cycle Arrest and Induction of Apoptosis

Cell cycle is mediated and controlled at different stages by cyclins, cyclin dependent kinases (CDKs), and cyclin dependent kinase inhibitors (CDKIs). Their expression and function are disturbed during cancer development and progression (Mahmoud et al. 2014). Genistein causes transcriptional up-regulation of p21 and p27 (CDK-interacting proteins) and CDKI, which leads to the cell cycle arrest and apoptosis in both androgen sensitive and androgen independent cells. Additionally, genistein reduces CDK4 expression and, in some level, inhibits CDK2, cyclin D1 and cyclin E action. Treatment with genistein and daidzein together was shown to increase p53 and reduce cyclin B1 protein expression (Mahmoud et al. 2014).

The Wnt/b-catenin pathway involves growth factors that control the cell growth, differentiation, and apoptosis. The activation of this pathway promotes the accumulation of b-catenin in the nucleus, resulting in the transcription of specific genes for cellular proliferation. Abnormal signaling stimulates cell division by b-catenin nuclear translocation which leads to the expression of other proteins important for cell division, including cyclin D1, c-myc, and Cox-2 (Lepri et al. 2014). Several reports exist concerning the reduction of cell growth due to the effect of isoflavones on the inactivation of Wnt pathway genes. Genistein decreased b-catenin gene expression, which can consequently lead to a decline in the levels of cytoplasmic b-catenin that is independent of the cytosolic degradation of b-catenin by the multi-protein complex (APC-GSK-3b-axin) (Lepri et al. 2014).

Another possible way of the flavonoids antiproliferative activity may be mediated by the inhibition of survivin, which is a member of the inhibitor of the apoptosis protein (IAP) family that has been implicated in the control of cell division. Survivin is considered to participate in the development and progression of cancer by suppressing apoptosis and regulating cell division (Lepri et al. 2014). In some investigations the mRNA levels for the anti-apoptotic proteins, survivin and protease activated receptor 2 (PAR-2) were down-regulated after treatment with genistein, whereas the mRNA levels for the pro-apoptotic protein elafin was up-regulated (Mahmoud et al. 2014).

One more phase in the control of cell proliferation is based on FOXO (fork head box O) genes that determine the production of important tumor suppressors - FOXO proteins. Phosphorylation by MAPK (mitogen-activated protein kinase) inhibits the function some FOXO family members. Current studies have shown that isoflavones can also alter the expression of FOXO family genes. Genistein and daidzein down-regulate MAPK related pathway genes transcription, which consequently prevents MAPK mediated phosphorylation of FOXO proteins and therefore could increase FOXO protein expression (Mahmoud et al. 2014). Pols (DNA-dependent DNA polymerases) and Topos are nuclear enzymes involved in different cell cycle phases. Pols are required for catalytic processes in DNA replication. Topos modify the DNA topology in the way necessary for the replication, transcription, recombination, and segregation of daughter chromosomes. Selective inhibitors of Pols and Topos can be potentially useful anticancer agents. Genistein adequately fits to a specific binding pocket on the Topo II protein surface. Genistein binds directly to Topo II and inhibits its activity. This reaction causes the cell cycle arrest at the G2/M phase (Mizushina et al. 2013). Isoflavones increase basal and chemotherapy-induced apoptosis. Genistein and daidzein support apoptosis by up-regulation of phosphatase and tensin homolog (PTEN). PTEN is a tumor suppressor gene that inhibits the anti-apoptotic Akt signaling pathway (Mahmoud et al. 2014). Isoflavones can induce quick disintegration of mitochondrial membranes which consequently leads to cell death (Sakamoto et al. 2010).

Furthermore, isoflavones were found to down-regulate the anti-apoptotic Bcl-2 protein and increase pro-apoptotic Bax mRNA and protein levels as well as boost activation and expression of caspase-3. Genistein can also initiate apoptosis through inhibition of the proteasome (regulates degradation of excess or damaged proteins by proteolysis), which stops metabolism of pro-growth cell cycle proteins (Mahmoud et al. 2014). Genistein was found to be an inducer of tumor cell differentiation, by down-regulating epidermal growth factor and ErbB2/Neu receptors in cancer cells (Choi et al. 2013).

It is well known that p53 presents one of the most important tumor suppressor genes with essential role in controlling cell proliferation. Studies showed that p53-p21wafl/cipl/cdc2 cross-talk co-regulates cell cycle at different stages. The G2/M checkpoint prevents cells from commencing mitosis if DNA is damaged anywhere.

Through phosphorylation, p53 induces the expression of p21waf/cipl and GADD45α (Zhang et al. 2013). As a result, cdc2 and cdc25A activity is inhibited, causing the detaining of G2/M phase cells. Studies proved that treatment with genistein induced G2/M cell proliferation arrest and caused apoptosis via the p53/ATM pathway. Genistein caused up-regulation of p53/p21wafl/cipl, GADD45α and down-regulation of cdc2 and cdc25A which leads to DNA damage. This mechanism as a response blocks the cell cycle and initiates programmed cell death (Zhang et al. 2013).

Studies have found the significant correlation of diet rich in phytoestrogens and low plasma concentration of two potent mitogens - insulin and IGF-1. On the other hand, a higher plasma level of IGF-1 binding proteins (IGFBP) was detected after isoflavone intake. Such proportion of mitogens in blood can reduce cancer risk. Therefore, some authors believe that phytoestrogen effects are mediated by modifying the action of these mitogens (Choi et al. 2013).

Inhibition Angiogenesis and Metastasis

Matrix metalloproteinases (MMPs) are proteolytic enzymes that can give cancer cells the ability to invade surrounding tissues by degrading the extracellular matrix. Genistein can reduce the angiogenic and metastatic potential of cancer cells. It can decrease the incidence of cancer cell invasion into the lymphatic and blood vessels and in that way prevent cancer metastasis (Sanjeev Banerjee et al. 2008). Genistein was found to inhibit the expression of c-erbB-2 and consequently reduce the secretion of MMPs. Besides, genistein can successfully lower the constitutively high levels of MMP-9 within T-lymphocytes. Genistein can also down-regulate the expression of MMP-9 (92 kDa type IV collagenase), MMP-2, protease M, uPAR, VEGF, neuropilin, TSP, BPGF, LPA, TGF-β, TSP-1, and PAR-2, which are all agents that support invasion and metastasis (Sanjeev Banerjee et al. 2008). It may also up-regulate TIMP-1 (tissue inhibitor of metalloproteinases) and some trypsin inhibitors (Choi et al. 2013). On the other hand, genistein promotes the growth of connective tissue by up-regulating the connective tissue growth factor and activation peptide. Furthermore, genistein can inhibit endothelial cell growth which could prevent abnormal vascularization of cancers. Current investigations also showed that genistein stimulates metastatic suppressor kangai-1 (KA11) gene expression (Sanjeev Banerjee et al. 2008).

It is well known that TGF-β is one of the most important factors that regulate cell proliferation. TGF-β is also involved in angiogenesis. Genistein was proven to inhibit TGF- β signaling as well as vascular endothelial growth factor 165, platelet-derived growth factor, tissue factor and urokinase plasminogen activator. On the other hand, after treatment with genistein an up-regulation of angiogenesis inhibitors (plasminogen activator inhibitor-1, endostatin, angiostatin, and thrombospondin-1) was noted. Genistein seems to be the strongest inhibitor of angiogenesis among all other isoflavones. Thus, it could be used for anti-metastatic therapy (Sanjeev Banerjee et al. 2008).

Table 4. Anticancer activity of isoflavones genistein and daidzein

Function	Target structure	Outcome
Down - regulation	ERα expression	hormone-related tumors growth reduction
	aromatase and 17β-oxidoreductase	inhibition of estrogen metabolism
	cyclin dependent kinases CDK4 and CDK2	cell-cycle arrest (G2/M) and induction of apoptosis
	cyclin B1, cyclin D1 and cyclin E	
	Wnt/b-catenin pathway genes and enzymes	
	MAPK (mitogen-activated protein kinase)	
	anti-apoptotic protein survivin	
	protease activated receptor 2 (PAR-2)	
	cell division regulator Topo II	
	anti-apoptotic Bcl-2 protein	
	proteasome for damaged proteins metabolism	
	mitogens insulin and IGF-1	
	metastatic suppressor kangai-1 (KA11)	
	c-erbB-2 / Neu	inhibition of angiogenesis and metastasis
	MMP (matrix metalloproteinases)	
	vascular endothelial growth factor 165	
	Pro-angiogenic TGF- β	
	urokinase plasminogen activator	
	platelet-derived growth factor	
	epidermal growth factor and ErbB2/Neu	reduction of oxidative stress
	membrane fluidity	
	erythrocyte cytosolic enzyme OGA	
Up - regulation	response on ERβ	hormone-related tumors growth reduction
	sex hormone binding globulin levels (SHBG)	reduction of sex hormone levels
	cell-cycle regulating proteins p21 and p27	cell-cycle arrest (G2/M) and induction of apoptosis
	cyclin dependent kinases inhibitors (CDKI)	
	pro-apoptotic protein elafin	
	tumor suppressors FOXO (fork head box O)	
	phosphatase for enzyme activity regulation	
	tumor suppressor tensin homolog (PTEN)	
	pro-apoptotic Bax mRNA	
	p53/p21waf1/cip1 and GADD45α	
	IGF-1 binding proteins (IGFBP)	
	TIMP-1 (tissue inhibitor of metalloproteinases)	inhibition of angiogenesis and metastasis
	trypsin and plasminogen activator inhibitors	
	connective tissue growth factor	
	endostatin, angiostatin, thrombospondin-1	
	DNA resistance to oxidative damage	reduction of oxidative stress
	Mn-SOD (manganese superoxide dismutase)	
	GPx (glutathione peroxidase) and reductase	
	catalase and superoxide dismutase	

Authors also found that genistein significantly reduces nuclear accumulation of hypoxia-inducible factor-1α in PC-3 cells. This factor regulates VEGF (vascular endothelial growth factor) transcription in response to hypoxia. Therefore, it was postulated that genistein may inhibit tumor angiogenesis by inhibition of VEGF-mediated autocrine and paracrine signaling pathways between tumor cells and vascular endothelial cells. (Sanjeev Banerjee et al. 2008).

ANTIOXIDATIVE AND ANTI-INFLAMMATORY PROPERTIES OF ISOFLAVONES

Oxidation reactions (transfer of electrons or hydrogen from a substance to an oxidizing agent) are everyday part of normal metabolism. However, during these reactions free radicals can emerge, which can cause damage or death to the cell.

Therefore, oxidative stress is considered as an important part of many illnesses, particularly carcinomas (Augusti et al. 2012). Antioxidants are molecules that inhibit the oxidation of other molecules by being oxidized themselves or by removing free radical intermediates and in that way inhibiting other oxidation reactions.

Isoflavones were proven to lower markers of oxidative stress (red blood cell membrane fluidity, activity of the red blood cell cytosolic enzyme OGA and lymphocyte DNA susceptibility to oxidative stress) and intensify DNA resistance to oxidative damage (Terzic et al. 2013). Literature data show that genistein can exert the antioxidant effect by the same mechanism like the estradiol which is well known antioxidant.

The activity of both estadiol and genistein is based on activation of estrogen receptors as well as the MAPK (mitogenic-activated protein kinase) and NFkB (transcription nuclear factor - kB) pathway (Erba et al. 2012). Activation of NFkB by natural estradiol subsequently enhances expression of antioxidant enzymes Mn-SOD (manganese superoxide dismutase) and GPx (glutathione peroxidase). Moreover, genistein stimulates a number of enzymes, like catalase, superoxide dismutase, glutathione peroxidase and reductase that are involved in oxidative reactions in organism (Choi et al. 2013).

Isoflavones genistein and daidzein are well-known not only as antioxidants, but also as chemo-preventive and anti-inflammatory agents. They could be involved in regulation of the immune response and in that way reduce prevalence of chronic health disorders of their regular consumers (Masilamani et al. 2012).

REFERENCES

Arnal, J. F., Douin-Echinard, V., Tremollières, F., Terrisse, A. D., Sié, P., Payrastre, B., Guery, J. C., Bayard, F., Gourdy, P. Understanding the controversy about hormonal replacement therapy: insights from estrogen effects on experimental and clinical atherosclerosis. *Arch. Mal. Coeur. Vaiss.* 2007;100:554-62.

Ascenzi, P., Bocedi, A., Marino, M. Structure-function relationship of estrogen receptor alpha and beta: impact on human health. *Mol. Aspects Med.* 2006;27:299-402.

Augusti, K. T., Jose, R., Sajitha, G. R., Augustine, P. A Rethinking on the Benefits and Drawbacks of Common Antioxidants and a Proposal to Look for the Antioxidants in Allium Products as Ideal Agents: A Review. *Ind. J. Clin. Biochem.* 2012;27:6-20.

Banerjee, S., Li, Y., Wang, Z., Sarkar, F. H. Multi-targeted therapy of cancer by genistein. *Cancer Lett.* 2008;269:226-42.

Bedell, S., Nachtigall, M., Naftolin, F. The pros and cons of plant estrogens for menopause. *J. Steroid Biochem. Mol. Biol.* 2014:139:225-36.

Bitto, A., Polito, F., Squadrito, F., Marini, H., D'Anna, R., Irrera, N., Minutoli, L., Granese, R., Altavilla, D. Genistein aglycone: a dual mode of action anti-osteoporotic soy isoflavone rebalancing bone turnover towards bone formation. *Curr. Med. Chem.* 2010;17:3007-18.

Bolego, C., Poli, A., Cignarella, A., Paoletti, R. Phytoestrogens: Pharmacological and Therapeutic Perspectives. *Curr. Drug Targets* 2003;4:77-87.

Bovee, T. F., Schoonen, W. G., Hamers, A. R., Bento, M. J., Peijnenburg, A. A. Screening of synthetic and plant-derived compounds for (anti)estrogenic and (anti)androgenic activities. *Anal. Bioanal. Chem.* 2008;390:1111-9.

Bowers, J. L., Tyulmenkov, V. V., Jernigan, S. C., Klinge, C. M. Resveratrol acts as a mixed agonist/antagonist for estrogen receptors alpha and beta. *Endocrinology* 2000; 141: 3657-67.

Carlson, S., Peng, N., Prasain, J. K., Wyss, J. M. Effects of botanical dietary supplements on cardiovascular, cognitive, and metabolic function in males and females. *Gend. Med.* 2008;5:S76-90.

Cederroth, C. R., Zimmermann, C., Nef, S. Soy, phytoestrogens and their impact on reproductive health. *Mol. Cell. Endocrinol.* 2012;355:192-200.

Chen, M. N., Lin, C. C., Liu, C. Efficacy of phytoestrogens for menopausal symptoms: a meta-analysis and systematic review. *Climacteric* 2014;1:1-10.

Choi, E. J., Kim, G. H. Antiproliferative activity of daidzein and genistein may be related to ERα/c-erbB-2 expression in human breast cancer cells. *Mol. Med. Rep.* 2013; 7:781-4.

Choi, J. S., Song, J. Effect of genistein on insulin resistance, renal lipid metabolism, and antioxidative activities in ovariectomized rats. *Nutrition* 2009;25:676-85.

Cornwell, T., Cohick, W., Raskin, I. Dietary phytoestrogens and health. *Phytochemistry* 2004;65:995-1016.

Dai, J., Li, Y., Zhou, H., Chen, J., Chen, M., Xiao, Z. Genistein Promotion of Osteogenic Differentiation through BMP2/SMAD5/RUNX2 Signaling. *Int. J. Biol. Sci.* 2013;9:1089-1098.

De Andrade, C. M., Bianchini, F. J., Rey, F. M., Fonseca, M. J., Toloi, M. R. Effects of an aglycone-rich biotransformed soybean extract in human endothelial cells. *Climacteric* 2014;20:1-5.

Dijsselbloem, N., Vanden Berghe, W., De Naeyer, A., Haegeman, G. Soy isoflavone phyto-pharmaceuticals in interleukin-6 affections. Multi-purpose nutraceuticals at the crossroad of hormone replacement, anti-cancer and anti-inflammatory therapy. *Biochem. Pharmacol.* 2004;68:1171-85.

Erba, D., Casiraghi, M. C., Martinez-Conesa, C., Goi, G., Massaccesi, L. Isoflavone supplementation reduces DNA oxidative damage and increases O-β-N-acetyl-D-glucosaminidase activity in healthy women. *Nutr. Res.* 2012;32:233-40.

Franke, A. A., Lai, J. F., Halm, B. M. Absorption, distribution, metabolism, and excretion of isoflavonoids after soy intake. *Arch. Biochem. Biophys.* 2014;559:24-8.

Fraser, L. R., Beyret, E., Milligan, S. R., Adeoya-Osiguwa, S. A. Effects of estrogenic xenobiotics on human and mouse spermatozoa. *Hum. Reprod.* 2006;21:1184-93.

Gleason, C. E., Carlsson, C. M., Barnet, J. H., Meade, S. A., Setchell, K. D., Atwood, C. S., Johnson, S. C., Ries, M. L., Asthana, S. A preliminary study of the safety, feasibility and cognitive efficacy of soy isoflavone supplements in older men and women. *Age Ageing* 2009;38:86-93.

Gottstein, N., Ewins, B. A., Eccleston, C., Hubbard, G. P., Kavanagh, I. C., Minihane, A. M., Weinberg, P. D., Rimbach, G. Effect of genistein and daidzein on platelet aggregation and monocyte and endothelial function. *Br. J. Nutr.* 2003;89:607-16.

Hill, D. A., Hill, S. R. Counseling patients about hormone therapy and alternatives for menopausal symptoms. *Am. Fam. Physician* 2010;82:801-7.

Jerome-Morais, A., Diamond, A. M., Wright, M. E. Dietary supplements and human health: for better or for worse? *Mol. Nutr. Food Res.* 2011;55:122-35.

Jiang, Q., Payton-Stewart, F., Elliott, S., Driver, J., Rhodes, L. V., Zhang, Q., Zheng, S., Bhatnagar, D., Boue, S. M., Collins-Burow, B. M., Sridhar, J., Stevens, C., McLachlan, J. A., Wiese, T. E., Burow, M. E., Wang, G. Effects of 7-O substitutions on estrogenic and anti-estrogenic activities of daidzein analogues in MCF-7 breast cancer cells. *J. Med. Chem.* 2010;53:6153-63.

Jungbauer, A., Medjakovic, S. Phytoestrogens and the metabolic syndrome. *J. Steroid Biochem. Mol. Biol.* 2014;139:277-89.

Kim, J. H., Kim, D., Kim, J., Hwang, J. K. Euchresta horsfieldii Benn. activates peroxisome proliferator-activated receptor α and regulates expression of genes involved in fatty acid metabolism in human HepG2 cells. *J. Ethnopharmacol.* 2011;133:244-7.

Ko, K. P. Isoflavones: chemistry, analysis, functions and effects on health and cancer. *Asian Pac. J. Cancer Prev.* 2014;15:7001-10.

Krum, S. A., Miranda-Carboni, G. A., Hauschka, P. V., Carroll, J. S., Lane, T. F., Freedman, L. P., Brown, M. Estrogen protects bone by inducing Fas ligand in osteoblasts to regulate osteoclast survival. *EMBO J.* 2008;27:535-45.

Kuhnle, G. G., Dell'aquila, C., Aspinall, S. M., Runswick, S. A., Mulligan, A. A., Bingham, S. A. Phytoestrogen content of cereals and cereal-based foods consumed in the UK. *Nutr. Cancer* 2009;61:302-9.

Kurzer, M. S. Hormonal effects of soy in premenopausal women and men. *J. Nutr.* 2002;132:S570-3.

Lagarib, V. S., Levis, S. Phytoestrogens for menopausal bone loss and climacteric symptoms. *J. Steroid Biochem. Mol. Biol.* 2014;139:294-301.

Lepri, S. R., Campos Zanelatto, L., Benites Goncalves da Silva, P., Sartori, D., Ribeiro, L. R., Mantovani, M. S. The effects of genistein and daidzein on cell proliferation kinetics in HT29 colon cancer cells: the expression of CTNNBIP1 (b-catenin), APC (adenomatous polyposis coli) and BIRC5 (survivin). *Human Cell* 2014;27:78-84.

Liao, Q. C., Xiao, Z. S., Qin, Y. F., Zhou, H. H. Genistein stimulates osteoblastic differentiation via p38 MAPK-Cbfa1 pathway in bone marrow culture. *Acta Pharmacol. Sin.* 2007;28:1597-602.

Liu, D., Homan, L. L., Dillon, J. S. Genistein acutely stimulates nitric oxide synthesis in vascular endothelial cells by a cyclic adenosine 5'-monophosphate-dependent mechanism. *Endocrinology* 2004;145:5532-9.

Lorand, T., Vigh, E., Garai, J. Hormonal action of plant derived and anthropogenic non-steroidal estrogenic compounds: phytoestrogens and xenoestrogens. *Curr. Med. Chem.* 2010;17:3542-74.

Luoto, R. Hot flushes and quality of life during menopause. *BMC Womens Health* 2009;9:13.

Maggiolini, M., Bonofiglio, D., Marsico, S., Panno, M. L., Cenni, B., Picard, D., Andò, S. Estrogen receptor alpha mediates the proliferative but not the cytotoxic dose-dependent effects of two major phytoestrogens on human breast cancer cells. *Mol. Pharmacol.* 2001;60:595-602.

Mahmoud, A. M., Yang, W., Bosland, M. C. Soy isoflavones and prostate cancer: A review of molecular mechanisms. *J. Steroid Biochem. Mol. Biol.* 2014;140:116-32.

Marini, H., Minutoli, L., Polito, F., Bitto, A., Altavilla, D., Atteritano, M., Gaudio, A., Mazzaferro, S., Frisina, A., Frisina, N., Lubrano, C., Bonaiuto, M., D'Anna, R., Cannata, M. L., Corrado, F., Cancellieri, F., Faraci, M., Marini, R., Adamo, E. B., Wilson, S., Squadrito, F. OPG and sRANKL serum concentrations in osteopenic, postmenopausal women after 2-year genistein administration. *J. Bone Miner. Res.* 2008;23:715-20.

Masilamani, M., Wei, J., Sampson, H. A. Regulation of the immune response by soybean isoflavones. *Immunol. Res.* 2012;54:95-110.

McCarty, M. F. Isoflavones made simple - genistein's agonist activity for the beta-type estrogen receptor mediates their health benefits. *Med. Hypotheses* 2006;66:1093-114.

Messina, M. Insights gained from 20 years of soy research. *J. Nutr.* 2010;140:2289-95.

Mizushina, Y., Shiomi, K., Kuriyama, I., Takahashi, Y., Yoshida, H. Inhibitory effects of a major soy isoflavone, genistein, on human DNA topoisomerase II activity and cancer cell proliferation. *Int. J. Oncol.* 2013;43:1117-24.

Molla, M. D., Hidalgo-Mora, J. J., Soteras, M. G. Phytotherapy as alternative to hormone replacement therapy. *Front. Biosci.* (Schol Ed.) 2011;3:191-204.

Moon, Y. J., Wang, X., Morris, M. E. Dietary flavonoids: effects on xenobiotic and carcinogen metabolism. *Toxicol. In Vitro* 2006;20:187-210.

Neill, A. S., Ibiebele, T. I., Lahmann, P. H., Hughes, M. C., Nagle, C. M., Webb, P. M., Australian Ovarian Cancer Study Group and Australian National Endometrial Cancer Study Group. Dietary phyto-oestrogens and the risk of ovarian and endometrial cancers: findings from two Australian case-control studies. *Br. J. Nutr.* 2014;111:1430-40.

Pérez-Jiménez, J., Hubert, J., Hooper, L., Cassidy, A., Manach, C., Williamson, G., Scalbert, A. Urinary metabolites as biomarkers of polyphenol intake in humans: a systematic review. *Am. J. Clin. Nutr.* 2010;92:801-9.

Polotsky, H. N., Polotsky, A. J. Metabolic implications of menopause. *Semin. Reprod. Med.* 2010;28:426-34.

Roberts, H., Lethaby, A. Phytoestrogens for menopausal vasomotor symptoms: A Cochrane review summary. *Maturitas* 2014;78:79-81.

Rozman, K. K., Bhatia, J., Calafat, A. M., Chambers, C., Culty, M., Etzel, R. A., Flaws, J. A., Hansen, D. K., Hoyer, P. B., Jeffery, E. H., Kesner, J. S., Marty, S., Thomas, J. A., Umbach, D. NTP-CERHR expert panel report on the reproductive and developmental toxicity of genistein. *Birth Defects Res. B Dev. Reprod. Toxicol.* 2006; 77:485-638.

Sakamoto, Y., Naka, A., Ohara, N., Kondo, K., Iida, K. Daidzein regulates proinflammatory adipokines thereby improving obesity-related inflammation through PPARγ. *Mol. Nutr. Food Res*. 2014;58:718-26.

Sanchez-Borrego, R., Navarro, M. C., Llaneza, P., Hormigo, A., Duran, M., Mendoza, N. Efficacy and safety of a phyto-SERM as an alternative to hormone therapy. *Climacteric* 2014;3:1-8.

Seidlova-Wuttke, D., Jarry, H., Wuttke, W. Plant derived alternatives for hormone replacement therapy (HRT). *Horm. Mol. Biol. Clin. Investig*. 2013;16:35-45.

Setchell, K. D., Clerici, C. Equol: history, chemistry, and formation. *J. Nutr*. 2010;140: 1355-62.

Setchell, K. D. R. Soy Isoflavones—Benefits and Risks from Nature's Selective Estrogen Receptor Modulators (SERMs). *J. Am. Coll. Nutrition* 2001;20:354-62.

Shen, J., Li, W., Liu, G., Tang, Y., Jiang, H. Computational insights into the mechanism of ligand unbinding and selectivity of estrogen receptors. *J. Phys. Chem. B* 2009;113:10436-44.

Sirotkin, A. V., Harrath, A. H. Phytoestrogens and their effects. *Eur. J. Pharmacol*. 2014;741:230-236.

Sirtori, C. R., Eberini, I., Arnoldi, A. Hypocholesterolaemic effects of soya proteins: results of recent studies are predictable from the Anderson meta-analysis data. *Br. J. Nutr*. 2007;97:816-22.

Somponpun, S., Sladek, C. D. Role of estrogen receptor-beta in regulation of vasopressin and oxytocin release in vitro. *Endocrinology* 2002;143:2899-904.

Soni, M., Rahardjo, T. B. W., Soekardi, R., Sulistyowati, Y., Yesufu-Udechuku, A., Irsan, A., Hogervorst, E. Phytoestrogens and cognitive function: a review. *Maturitas* 2014;77:209-20.

Staar, S., Richter, D. U., Makovitzky, J., Briese, V., Bergemann, C. Stimulation of endometrial glandular cells with genistein and daidzein and their effects on ERalpha- and ERbeta-mRNA and protein expresion. *Anticancer Res*. 2005;25:1713-8.

Strom, B. L., Schinnar, R., Ziegler, E. E., Barnhart, K. T., Sammel, M. D., Macones, G. A., Stallings, V. A., Drulis, J. M., Nelson, S. E., Hanson, S. A. Exposure to soy-based formula in infancy and endocrinological and reproductive outcomes in young adulthood. *JAMA* 2001;286:807-14.

Terzic, M., Dotlic, J., Maricic, S., Mihailovic, T., Tosic-Race, B. The influence of red clover derived isoflavones on serum lipid profile in postmenopausal women. *J. Obstet. Gynecol. Res*. 2009;35:1091-5.

Terzic, M., Dotlic, J. Phytoestrogens in gynecology - antioxidative role. *Oxid. Antioxid. Med. Sci*. 2013;2:1-4.

Terzic, M., Micic, J., Dotlic, J., Maricic, S., Mihailovic, T., Knezevic, N. Impact of Phytoestrogens on Serum Lipids in Postmenopausal Women. *Geburtshilfe Frauenheilkd* 2012;72:527-31.

Terzic, M., Micic, J., Dotlic, J. Phytoestrogen Use in Menopausal Patients: Current Clinical Approach. In: Fiala, J., Pospisil, D. (Eds.). *Alfalfa and Clovers: Properties, Medicinal Uses and Health Benefits*. Nova Science Publishers, Inc, New York, US, 2012,75-93.

Thorp, A. A., Howe, P. R., Mori, T. A., Coates, A. M., Buckley, J. D., Hodgson, J., Mansour, J., Meyer, B. J. Soy food consumption does not lower LDL cholesterol in either equol or nonequol producers. *Am. J. Clin. Nutr*. 2008;88:298-304.

Van der Woude, H., Ter Veld, M. G., Jacobs, N., van der Saag, P. T., Murk, A. J., Rietjens, I. M. The stimulation of cell proliferation by quercetin is mediated by the estrogen receptor. *Mol. Nutr. Food Res*. 2005;49:763-71.

Vilar-Gonzalez, S., Perez-Rozos, A., Cabanillas-Farpon, R. Mechanism of hot flashes. *Clin. Transl. Oncol*. 2011;13:143-7.

Vogelvang, T. E., van der Mooren, M. J., Mijatovic, V., Kenemans, P. Emerging selective estrogen receptor modulators: special focus on effects on coronary heart disease in postmenopausal women. *Drugs* 2006;66:191-221.

Wang, H., Li, J., Gao, Y., Xu, Y., Pan, Y., Tsuji, I., Sun, Z. J., Li, X. M. Xeno-oestrogens and phyto-oestrogens are alternative ligands for the androgen receptor. *Asian J. Androl*. 2010;12:535-47.

Yuan, J. P., Wang, J. H., Liu, X. Metabolism of dietary soy isoflavones to equol by human intestinal microflora--implications for health. *Mol. Nutr. Food Res*. 2007;51: 765-81.

Zhang, Z., Wang, C. Z., Du, G. J., Qi, L. W., Calway, T., He, T. C., Du, W., Yuan, C. S.Genistein induces G2/M cell cycle arrest and apoptosis via ATM/p53-dependent pathway in human colon cancer cells. *Int. J. Oncol*. 2013;43:289-96.

Zhao, C., Matthews, J., Tujague, M., Wan, J., Ström, A., Toresson, G., Lam, E. W., Cheng, G., Gustafsson, J. A., Dahlman-Wright, K. Estrogen receptor beta2 negatively regulates the transactivation of estrogen receptor alpha in human breast cancer cells. *Cancer Res*. 2007;67:3955-62.

Zheng, W., Hou, Y., Yao, W. Equol: a metabolite of soy isoflavones by intestinal microflora - a review. *Wei Sheng Wu Xue Bao* 2013;53:1251-7.

In: Genistein and Daidzein
Editor: Masayoshi Yamaguchi
ISBN: 978-1-63483-194-9

Chapter 5

BONE FORMATION: ROLES OF GENISTEIN AND DAIDZEIN

Marjorie E. Bateman[1,*], Amy L. Strong[1], Matthew E. Burow[2], Guangdi Wang[3], Stephen M. Boue[4] and Bruce A. Bunnell[1]
[1]Center for Stem Cell Research and Regenerative Medicine, Tulane University School of Medicine, New Orleans, Louisiana, US
[2]Department of Medicine, Tulane University School of Medicine, New Orleans, Louisiana, US
[3]Department of Chemistry and RCMI Cancer Research Center, Xavier University, New Orleans, Louisiana, US
[4]Southern Regional Research Center, US Department of Agriculture, New Orleans, Louisiana, US

ABSTRACT

Bone remodeling consists of a balance between bone formation by osteoblasts and bone resorption by osteoclasts. Osteoporosis is the result of increased bone resorption and decreased bone formation causing a decreased bone mass density, loss of bone microarchitecture, and an increased risk of fractures. Osteoporosis is most common in postmenopausal women, primarily due to the effects of hypoestrogenemia. In this chapter, the efficacy and mechanisms of action of phytoestrogens in promoting osteogenesis and the use of phytoestrogens as potential therapeutic agents for osteoporosis will be presented. Two phytoestrogens that have gained significant attention are daidzein and genistein. Overall, daidzein and genistein share many cellular effects to reverse the phenotypes observed in ovariectomized animals. Studies have shown that genistein effectively reduces the impacts of osteoporosis, while efficacy studies investigating daidzein have shown mixed results, which will be discussed. Several studies showed that daidzein is more efficient than genistein in preventing ovariectomy-induced bone loss in rats. With regard to the molecular signaling that induces these phenotypic cell-specific effects, daidzein and genistein both act primarily through estrogen receptor

[*] Corresponding author:Bruce Bunnell. Center for Stem Cell Research and Regenerative Medicine, Tulane University School of Medicine, New Orleans, Louisiana 70112, US. E-mail: bbunnell@tulane.edu.

dependent mechanisms at lower concentrations but may also act through different pathways at higher concentrations, which will be presented in this chapter. The importance of metabolites of daidzein and genistein and other phytoestrogens that may play a role in osteogenesis will also be addressed.

ABBREVIATIONS

ALP	alkaline phosphatase
AP-1	activator protein-1
ASCs	adipose-derived stem cells
BMD	bone mineral density
BMP	bone morphogenetic protein
BMSCs	bone marrow-derived stem cells
BSP	bone sialoprotein
Cbfa1	core binding factor 1
cGMP	cyclic guanosine monophosphate
CREB	cAMP response element-binding protein
DC-STAMP	dendritic cell-specific transmembrane protein
DHP	dihydropyridine
DKK1	dickkopf
EC50	half-maximal effective concentration
EGFR	epidermal growth factor receptor
ER	estrogen receptor
ERα	estrogen receptor alpha
ERβ	estrogen receptor beta
ERK 1/2	extracellular signal-regulated kinase 1/2
FOS	fructooligosaccharides
HO-1	heme oxygenase-1
hOB	human osteoblast cells
IGF1	insulin-like growth factor 1
IKir	inward rectifier potassium current
IL-1	interleukin-1
IL-1β	interleukin-1 beta
IL-6	interleukin-6
JNK 1/2	c-Jun N-terminal kinase 1/2
MAPK	mitogen-activated protein kinase
M-CSF	macrophage colony-stimulating factor
MSCs	mesenchymal stem cells
NADPH	nicotinamide adenine dinucleotide phosphate
NFATc1	nuclear factor of activated T cells
NF-κB	nuclear factor-kappaB
NO	nitric oxide
Nox1	nicotinamide adenine dinucleotide phosphate oxidase
OC	osteocalcin
OPG	osteoprotegerin

OPN	osteopontin
ORX	orchidectomized
OVX	ovariectomized
PGE2	prostaglandin E2
PKA	protein kinase A
PKC	protein kinase C
PPARδ	peroxisome proliferator activated receptor delta
PPARγ	peroxisome proliferator activated receptor gamma
PTH	parathyroid hormone
PTHR	parathyroid hormone receptor
RANK	receptor activator of nuclear kappa B
RANKL	receptor activator of nuclear factor-kappa B ligand
ROS	reactive oxygen species
SERM	selective estrogen receptor modulator
SOD	superoxide dismutase
TNF-α	tumor necrosis factor-alpha
TRAF2	tumor necrosis factor receptor associated factor 2
TRAF3	tumor necrosis factor receptor associated factor 3
TRAF6	tumor necrosis factor receptor associated factor 6
VDR	vitamin D receptor

1. Osteoporosis

1.1. Epidemiology of Osteoporosis

Osteoporosis is a pathological condition characterized by reduced bone strength and increased susceptibility to fractures, which results in significant morbidity and mortality (Bliuc et al., 2009) (Bliuc et al., 2015) (Nazrun, Tzar, Mokhtar, and Mohamed, 2014) (Sattui and Saag, 2014).

According to the National Center for Health Statistics, nine percent of adults 50 years of age or older suffered from osteoporosis at either the femur neck or lumbar spine in the United States from 2005-2008 (Looker AC, 2012). In 2005, there were greater than two million osteoporosis-related fractures in the United States, which cost nearly $17 billion (Burge et al., 2007).

In addition to being a prevalent disease, osteoporosis is a growing problem. While there are many risk factors associated with the development of osteoporosis, the most significant risk factor for bone loss in men and women is aging (Schroder, 2015). It has been demonstrated that both men and women have a continuous slow phase of age-related bone loss occurring at 1% per year (Marie and Kassem, 2011). The aging of the United States population is, therefore, likely to lead to increased prevalence of osteoporosis. It is predicted that in 2025 there will be greater than three million fractures with a cost of approximately $25.3 billion (Burge et al., 2007).

1.2. Osteoporosis in Men

In the United States, prevalence of osteoporosis in men has been shown to be 3-4% until 80 years of age when it increases to 10% (Looker AC, 2012). Determining the pathways associated with osteoporosis and efficacy of potential therapies in men is important as men account for greater than twenty-five percent of the economic burden of osteoporosis (Burge et al., 2007). Osteoporosis in men can be classified as primary (age-related) or secondary (caused by hypogonadism, chronic diseases, drug therapy, or lifestyle practices that lead to bone loss). Androgens and estrogens are involved in skeletal homeostasis, and loss of bone in men is associated with significant reductions of bioavailable serum testosterone and estrogen with advancing age (Stathopoulos, Ballas, Lampropoulou-Adamidou, and Trovas, 2014) (Sinnesael, Claessens, Boonen, and Vanderschueren, 2013).

1.3. Osteoporosis in Women

In the United States, the prevalence of osteoporosis in women at either the femur neck or lumbar spine was 7% for 50-59 year olds, 10% for 60-69 year olds, 27% for 70-79 year olds, and 35% for 80-89 year olds (Looker AC, 2012). In addition to the continuous slow phase of age-related bone loss occurring at 1% per year that men experience, perimenopausal women experience a rapid drop in estrogen levels leading to bone loss of 5-10% per year (Marie and Kassem, 2011). Bone fragility is, therefore, more common in women than in men partly because the production of sex hormones in males does not decrease rapidly and because there is no rapid midlife increase in the rate of bone remodeling (Schroder, 2015).

2. Cellular Changes in Osteoporosis

2.1. Cortical and Trabecular Bone

The two types of bone are cortical bone and trabecular bone. Cortical bone is used to build long bones, so it makes sense that cortical bone would favor rigidity over flexibility for functions of movement and loading.

In contrast, trabecular bone in the vertebral bodies is assembled as an open-celled, porous structure with interconnecting trabecular plates. Trabecular bone favors structural flexibility over stiffness, which allows for some degree of deformation prior to cracking (Seeman and Delmas, 2006).

2.2. Normal Bone Remodeling

Osteoclasts and osteoblasts are the cells involved in normal bone remodeling that become dysregulated in pathologic conditions such as osteoporosis. Osteoclasts are cells derived from precursors in the myeloid/ monocyte lineage in the bone marrow (Boyce, Yao, and Xing, 2009).

These osteoclast precursors fuse to form multinucleated osteoclastic cells that secrete hydrochloric acid and proteases, such as cathepsin K, to dissolve mineral and matrix components of bone (Boyce et al., 2009). Osteoclast differentiation is regulated by cells called osteoblasts, which are derived from mesenchymal stem cells (MSCs) in the bone marrow (Capulli, Paone, and Rucci, 2014). Osteoblasts produce factors that can positively regulate osteoclast differentiation and also factors that can inhibit osteoclast differentiation, depending on the physiological context (Capulli et al., 2014). Osteoblasts are primarily known for their bone building function, which involves deposition of an organic matrix and its subsequent mineralization (Capulli et al., 2014).

Normal bone remodeling begins with the activation phase in which osteoclasts are recruited to the surface of the bone. In the resorption phase, osteoclasts generate an acidic microenvironment between the cell and the surface of the bone to aid in resorption of the mineral content in the bone. In the reversal phase, osteoclasts undergo apoptosis and osteoblasts are recruited to the surface of the bone. To complete remodeling of bone, osteoblasts deposit collagen which is then mineralized to form new bone in the formation phase (Tella and Gallagher, 2014).

2.3. Pathologic Bone Remodeling Associated with Aging

The key to the common pathway between men and women in age-related bone loss is the pathologic remodeling of cortical and trabecular bone. Remodeling occurs on bone surfaces, and due to the larger surface area in trabecular bone compared to cortical bone, there are more remodeling sites per unit volume in trabecular bone (Seeman and Delmas, 2006). During aging, cellular changes in osteoblasts and osteoclasts cause each remodeling event to remove more bone than it replaces, which means that a greater proportion of trabecular bone is turned over and removed. The high remodeling rate and deep resorption cavities produce a loss of trabecular plates and their connectivity which produce a deficit in bone strength. Rapid remodeling persists with aging due to hypogonadism in some men and women. Both men and women may experience secondary hyperparathyroidism which results in age-related bone loss. With continued remodeling, trabeculae perforate and some disappear entirely. Remodeling then becomes more active on endocortical surfaces (endosteal bone resorption) than on remaining trabecular surfaces, and bone loss becomes primarily cortical in origin. Increased cortical thinning and porosity leads to crack formation and spread.

As pores coalesce, the reduced bone mass is unable to absorb energy imparted by a fall, resulting in increased susceptibility to fractures (Seeman and Delmas, 2006).

2.4. Comparison of Aging-Related Bone Loss in Men and Women

It is important to note the differences in pathology of aging-related bone loss between men and women as these distinctions may be relevant for developing more specific treatments. With regard to trabecular bone, the majority of bone loss in aging men can be attributed to trabecular thinning from reduced bone formation whereas a minority of bone loss in men is attributed to trabecular perforations and loss of trabecular connectivity (Marie and Kassem, 2011).

Bone loss in postmenopausal women occurs primarily by reduced number of trabeculae and increased trabecular perforation, a process related to increased bone resorption, not just reduced bone formation. This fact may explain increased trabecular bone fragility and earlier increase in fracture risk in women (Marie and Kassem, 2011).

There are also differences between men and women with regard to cortical bone loss. Cortical bone mass remains stable until women reach menopause and then exhibits a greater age-related decrease in women than in men (Marie and Kassem, 2011).

Androgens and estrogens are needed for skeletal homeostasis in both men and women. Estrogen deficiency is a primary determinant of age-related bone loss in both sexes (Marie and Kassem, 2011). While estrogen deficiency is a contributor during the continuous phase of bone loss in aging men, it results in accelerating the early, slow phases of bone loss in postmenopausal women (Sinnesael et al., 2013).

Therefore, estrogen deficiency in postmenopausal women and elderly men may result in changes at the cellular level that lead to the accelerated bone remodeling with bone resorption exceeding bone formation (Marie and Kassem, 2011).

2.5. Cellular Changes in Aging-Related Bone Loss

The critical pathogenic mechanism associated with aging-related skeletal fragility is impaired bone formation due to insufficient number of osteoblasts (Almeida, 2012) (Marie and Kassem, 2011) (Manolagas, 2014). This decreased osteoblast number may be from reduction in life-span of osteoblasts, which is enhanced by estrogen deficiency and androgen deficiency (Marie and Kassem, 2011) (Seeman and Delmas, 2006) (Manolagas, 2014). Decreased osteoblast number has also been attributed to decreased MSCs, defective proliferation and differentiation of these progenitor cells, or redirection of these progenitors toward adipocyte lineage (Almeida, 2012) (Hu et al., 2013).

In addition to affecting osteoblasts, estrogen deficiency increases the rate of remodeling and the volume of bone resorbed by prolonging the life span of osteoclasts and increasing osteoclastic resorption activity (Seeman and Delmas, 2006) (Tella and Gallagher, 2014). Some of these effects on osteoclasts are mediated through an increase in the production of proinflammatory cytokines with the onset of menopause. Interleukin-1 (IL-1), interleukin-6 (IL-6), and tumor necrosis factor-alpha (TNF-α) levels have been demonstrated to be increased in patients with osteoporosis, and in postmenopausal women, these factors have been shown to be associated with bone mineral density (BMD) (McLean, 2009). Under conditions of estrogen deficiency, IL-1 and TNF-α stimulate release of IL-6, macrophage colony-stimulating factor (M-CSF), and other proinflammatory cytokines by stromal cells/ preosteoblasts, resulting in further insult to bone integrity (Tella and Gallagher, 2014).

These proinflammatory cytokines have a variety of effects which promote bone resorption. IL-1, IL-6, and TNF-α have been shown to increase osteoclast differentiation, activation and survival, promote receptor activator of nuclear factor-kappa B ligand (RANKL) expression, and inhibit the survival of osteoblasts (McLean, 2009). Studies have demonstrated significant negative correlations between IL-1β, IL-6, and TNF-α secretion and lumbar spine BMD (McLean, 2009), and IL-6 and TNF-α have been shown to be associated with a high reduction in bone mineral content (BMC) and BMD in postmenopausal women across a variety of bone sites, particularly the hip (Gertz et al., 2010).

IL-6 may be the most influential cytokine in bone loss in estrogen deficiency in the early postmenopausal stage since serum IL-6 has been demonstrated to have a negative correlation with BMD at the hip and the wrist (McLean, 2009) (Rickard et al., 2003). Recent studies provide evidence that increased TNF-α levels may be an important risk factor for fracture (McLean, 2009). Furthermore, TNF-α knockout mice do not experience rapid decrease in BMD post ovariectomy, which suggests that TNF-α may be a key player in the accelerated bone loss that occurs following onset of menopause (Weitzmann and Pacifici, 2006a) (Weitzmann and Pacifici, 2006b). Additional studies have shown that TNF-α enhances stromal cell production of IL-6, M-CSF, and RANKL, which are essential stimuli of osteoclastogenesis and may result in increasing bone resorption in postmenopausal women (Weitzmann and Pacifici, 2006b) (Weitzmann and Pacifici, 2006a) (Tella and Gallagher, 2014). Estrogen deficiency also stimulates osteoclastogenesis by failure to inhibit the RANKL/RANK signaling pathway. Estrogen increases osteoprotegerin (OPG) secretion by osteoblast lineage cells and decreases expression of M-CSF and receptor activator of nuclear kappa B (Heim et al.), which increases inhibition of the RANKL/RANKL pathway (Tella and Gallagher, 2014). Inhibition of the RANKL/RANK signaling pathway prevents induction of expression of genes that cause preosteoclasts to differentiate into mature osteoclasts.

In addition to changes in osteoblasts and osteoclasts that favor resorption, B and T lymphocytes also contribute to bone loss in osteoporotic patients. Estrogen regulates B and T cell functions and their interactions with osteoclasts and osteoblasts in bone (Weitzmann and Pacifici, 2006b). Estrogen deficiency results in the expansion of B cell populations (Weitzmann, 2014). Cytokine production and increased RANKL production by B and T lymphocytes contribute to bone loss in postmenopausal women (Onal et al., 2012) (Weitzmann, 2014) (Boyle, Simonet, and Lacey, 2003). T cells from osteoporotic, estrogen-deficient postmenopausal patients also demonstrated increased secretion of TNF-α, which is a cytokine shown to play a role in increasing resorption (Weitzmann and Pacifici, 2006b) (Weitzmann and Pacifici, 2006a) (Weitzmann, 2014).

Age-related decline in bone mass is also associated with a progressive increase in oxidative stress (Manolagas, 2014) (Schroder, 2015) (Almeida, 2012). Estrogen and androgen deficiency also accelerates age-associated oxidative stress (Marie and Kassem, 2011) (Almeida, 2012) (Schroder, 2015). Decreased antioxidative capacity and increased reactive oxygen species (ROS) formation, together with TNF-α, may be at least partially responsible for aging and estrogen-depleted osteoporosis (Schroder, 2015).

3. Current Treatments for Osteoporosis

Drugs that interfere with the resorptive pathway or that amplify steps in the anabolic pathway are needed to treat osteoporosis and to prevent fractures associated with osteoporosis. Current therapies are primarily antiresorptive agents which prevent or slow the progression of bone fragility, but it remains necessary to restore bone lost due to remodeling to reestablish normal bone architecture and increase strength. Antiresorptive agents include bisphosphonates, estrogens (with and without progestins), selective estrogen receptor modulators (SERMs), tissue-specific estrogen complex (Bazedoxifene-estrogen), denosumab, cathepsin K inhibitors, and integrin antagonists (Tella and Gallagher, 2014).

Data from randomized, controlled trials suggest that bisphosphonates reduce the risk of vertebral fractures (Black, Bauer, Schwartz, Cummings, and Rosen, 2012). Estrogen has both anti-resorptive activity and anabolic activity and has been shown to reduce development of osteoporosis and increase BMD. However, its use in hormone replacement therapy has declined due to the association with increased risk of breast and endometrial cancer (Schairer et al., 2000) (Colditz et al., 1995) (Lambrinoudaki, 2014).

Similarly, estrogens have been shown to have a protective effect on bone density and metabolism in male animals, but estrogen replacement therapy is not feasible in men because of potential side effects (Filipovic et al., 2010). Raloxifene, a SERM, has been shown to produce estrogen agonistic effects on bone and estrogen-antagonistic effects on uterine, endometrium, and breast tissue (Fuchs-Young et al., 1995). However, raloxifene has also been associated with increased risk of venous thromboembolic events and fatal stroke (Barrett-Connor et al., 2006). Denosumab is a human monoclonal antibody approved for treatment of osteoporosis that specifically binds to RANKL and blocks the binding of RANK to its ligand. This binding has been shown to reduce bone resorption, increase bone density, and reduce fractures (Tella and Gallagher, 2014). Cathepsin K is a proteolytic enzyme known to degrade bone matrix, and cathepsin K inhibitors such as odanacatib show increasing bone density and decreased fractures (Tella and Gallagher, 2014).

Anabolic agents include teriparatide, anti-sclerostin antibodies, and dickkopf (DKK1) antibodies (Tella and Gallagher, 2014). Of the anabolic agents, only teriparatide, parathyroid hormone (PTH 1-34), is currently approved for the treatment of osteoporosis. Teriparatide stimulates bone formation by inhibiting sclerostin, DKK1, and frizzled protein. Sclerostin produced by osteocytes and DKK1 protein secreted in bone have been shown to be negative regulators of the Wnt signaling bone formation pathway, a pathway which is key for enhancing the MSCs commitment to and differentiation along the osteoblastic lineage (Tella and Gallagher, 2014) (Rossini, Gatti, and Adami, 2013) (Manolagas, 2014) (Baron and Kneissel, 2013). Thus, PTH 1-34 inhibition of sclerostin, DKK1, and frizzled protein increases BMD and improves microarchitecture (Tella and Gallagher, 2014). It has been shown to significantly decrease vertebral fractures in patients with severe osteoporosis but with only modest effect on nonvertebral fractures (Montagnani, 2014). Anti-sclerostin antibodies, such as romosozumab and blosozumab, have also been shown to increase bone mass (Tella and Gallagher, 2014). Recently, phytoestrogens such as genistein and daidzein have drawn significant attention as these compounds have been shown to increase osteogenesis through estrogen signaling pathways, without increasing cancer risk (Gercel-Taylor, Feitelson, and Taylor, 2004) (Jin, Zhang, Kang, Wang, and Zhao, 2010) (Ju, Fultz, Allred, Doerge, and Helferich, 2006) (Lamartiniere, Wang, Smith-Johnson, and Eltoum, 2002) (X. Liu, Suzuki, Santosh Laxmi, Okamoto, and Shibutani, 2012). The efficacy and mechanisms of action of daidzein and genistein are described below, and an overview of the cellular effects of genistein and daidzein is shown in Figure 1. It should be noted that BMD measurement was selected as a primary point of comparison of animal and human studies of isoflavone intervention efficacy. BMD measurement continues to be the most useful clinical tool for identifying patients with osteoporosis (Small, 2005), and it was the most commonly measured parameter, allowing comparisons to be drawn more readily between studies with regard to efficacy. However, bone strength and susceptibility to fracture also depend on trabecular connectivity and arrangement, biomechanical properties, and other factors, including bone size, shape, turnover, and architecture (Small, 2005).

When possible, the effects of isoflavones on these other parameters were included in this chapter for a more complete representation of the effects of isoflavones on bone strength and susceptibility to fracture. Both parameters are pertinent to whether isoflavones may represent a potential therapy for osteoporosis.

4. GENISTEIN

4.1. Efficacy Studies

4.1.A. Intervention in Animals

Ovariectomized (OVX) rat models are considered the most appropriate model for studying human menopausal osteoporosis due to similarities in pathophysiological mechanisms of bone deterioration (Kalu, 1991). OVX rats demonstrate sex hormone deficiency, leading to the accelerated loss of bone that mimics the rapid bone loss that women experience following menopause (Kalu, 1991). Shared characteristics between ovariectomy-induced bone loss in rats and postmenopausal bone loss in humans include increased rate of bone turnover with more resorption than formation, an initial rapid phase of bone loss followed by a slower phase, greater loss of trabecular bone than cortical bone, decreased intestinal absorption of calcium, and similar skeletal response to therapies such as estrogens, tamoxifen, bisphosphonates, parathyroid hormone, calcitonin, and exercise (Kalu, 1991).

Studies examining dietary administration of genistein alone at varying concentrations in OVX rats are noted in Table 2. Many studies demonstrated that genistein treatment resulted in the preservation of BMD in the whole femur (Nakajima et al., 2001) (Yang, Yu, Qu, and Li, 2014) (Z. L. Wang et al., 2006) (Bitto et al., 2008) (B. Li and Yu, 2003) (Bitto et al., 2011) (Miao et al., 2012) (Mathey et al., 2007), metaphyseal femur (Mathey et al., 2007), diaphyseal femur (Picherit et al., 2000), whole tibia (Fanti et al., 1998) (Z. L. Wang et al., 2006) (Hertrampf et al., 2007) (Hertrampf, Seibel, Laudenbach, Fritzemeier, and Diel, 2008) (Hertrampf et al., 2009) (Miao et al., 2012) and mandibular subchondral bone (Y. Q. Li et al., 2012). Genistein has been demonstrated to have an especially significant effect on trabecular bone in OVX rats, preserving trabecular bone BMD (Hertrampf et al., 2007) (Hertrampf et al., 2008) (Hertrampf et al., 2009) (Anderson, Ambrose, and Garner, 1998), trabecular thickness (Y. Q. Li et al., 2012) (Z. L. Wang et al., 2006) (Yang et al., 2014) (Mathey et al., 2007), trabecular area (Z. L. Wang et al., 2006) (Yang et al., 2014) (Mathey et al., 2007), trabecular number (Z. L. Wang et al., 2006) (Yang et al., 2014) (Mathey et al., 2007) (Anderson et al., 1998). In one study, genistein preserved the biomechanical quality of the trabecular bone, even though, vertebral trabecular microstructure and BMD was not preserved (R. Dai et al., 2008). However, other studies were unable to validate that genistein improved bone mineralization (BMD) in the fourth lumbar vertebrae (Nakajima et al., 2001) (Sliwinski et al., 2009), total vertebral trabecular bone (R. Dai et al., 2008), the total tibia (Sliwinski et al., 2009) (Turner et al., 2013), the proximal metaphysis of the tibia (Dodge et al., 1996), or the femoral diaphysis (Mathey et al., 2007). Therefore, in OVX rat models, the ability of genistein to preserve BMD was most consistently demonstrated in whole femur and whole tibia. Genistein also had positive effects on other trabecular bone parameters which demonstrated its ability to preserve the quality of trabecular bone.

Another model for postmenopausal osteoporosis is the OVX mouse. Studies examining dietary administration of genistein alone at varying concentrations in OVX mice are noted in Table 2. Studies conducted in OVX mice showed that genistein preserved BMD in the femur distal metaphysis (Erlandsson, Islander, Moverare, Ohlsson, and Carlsten, 2005) (Ishimi et al., 1999), whole femur (Power, Ward, Chen, Saarinen, and Thompson, 2006) (Wu et al., 2001) (Pie et al., 2006), and metaphyseal tibia (Sehmisch et al., 2010). One study reported that genistein supplementation resulted in superior biomechanical parameters relative to control and an improvement in trabecular connectivity, which is important for bone stability (Sehmisch et al., 2010). Treatment with genistein has been shown to play a role in preserving cancellous bone density (Sehmisch et al., 2010) (Cano et al., 2008) (Ishimi et al., 1999). Ishimi et al. demonstrated that treatment with genistein completely prevented ovariectomy-induced decrease in trabecular thickness and decreased ovariectomy-induced increase in trabecular spacing in mice (Ishimi et al., 1999). However, Cano et al. demonstrated that genistein was unable to prevent OVX-induced decrease in total volumetric BMD or microstructural deterioration (decreased trabecular number, decreased trabecular thickness, increased trabecular spacing, increased trabecular bone pattern factor, decreased trabecular bone volume) induced by ovariectomy in trabecular bone (Cano et al., 2008) (Zhang, Li, Wan, Helferich, and Wong, 2009). While some of these results are inconsistent and contraindicate the overall positive trend toward preventing bone loss seen in OVX rats, these results may in part be attributed to the fact that the OVX mouse model is not as accurate as the OVX rat model with regard to demonstrating the effects of genistein post-ovariectomy. After ovariectomy in mice, decreases in cancellous bone and greater bone turnover are not as high as those seen in estrogen-depleted rats or in osteoporotic women, which limits the usefulness of OVX mice as preclinical models (Sehmisch et al., 2010). Additionally, mice do not respond to exogenous estrogens in a manner similar to women (Turner et al., 2013).

To study effects of genistein in male animal models, orchidectomized (ORX) mice have been used. Genistein administration demonstrated significant suppression of orchidectomy-induced reduction in distal and whole femoral BMD (Wu et al., 2003) (Ishimi et al., 2002). Genistein at 0.4 mg/day is the minimal dose sufficient for bone-protective effect, in ORX mice (Wu et al., 2003). Genistein can also prevent trabecular bone volume loss in femoral distal metaphysis after orchidectomy (Ishimi et al., 2002).

4.1.B. Intervention in Humans

Genistein has been demonstrated to result in an increase in BMD in postmenopausal women at the lumbar spine and the femoral neck (Marini et al., 2007) (Morabito et al., 2002) (Crisafulli et al., 2004) (Atteritano et al., 2009). These results suggest that genistein produced a net gain in bone mass after one and two years of therapy (Marini et al., 2007).

Overall, based on animal studies, genistein seems to exert its bone protective effects on the whole femur and tibia and their metaphyseal zones. However, in human studies, genistein appears to exert bone protective effects on the lumbar spine and femoral neck. Interestingly, the distal femur and tibial metaphyseal zones are rich in cancellous bone as are the second, third, fourth, and fifth lumbar vertebrae.

As will be discussed in the following sections, genistein has higher affinity for estrogen receptor beta (ERβ) than for estrogen receptor alpha (ERα), and ERβ mRNA is found at higher levels in cancellous bone such as the distal femoral metaphysis, tibial metaphysis, and lumbar vertebrae, whereas ERα mRNA is predominantly expressed in the cortical bone of the femoral diaphysis (Bord, Horner, Beavan, and Compston, 2001).

Because the bone protection mechanisms of genistein are highly ER-dependent, it makes sense that genistein may be acting through ERβ in cancellous bone sites, such as the femoral and tibial metaphyseal sites and the lumbar vertebrae.

Dietary studies of efficacy of genistein intervention in humans and animals are summarized in Table 1 and Table 2, respectively. For animal dosage conversion to equivalent human doses, the Reagan-Shaw et al. study formula was used (Reagan-Shaw, Nihal, and Ahmad, 2008). It is difficult to draw comparisons between studies due to differences in study parameters such as age/menopausal stage of participants, duration of treatment, dietary habits of subjects, dosage of genistein, product form of genistein (pure compound in tablet or capsule, soy extracts, isoflavone enriched soy protein), storage and processing of isoflavone, size of treatment groups, and bone site and parameters evaluated. However, overall, genistein seems to have an effect to preserve BMD and bone quality at a variety of sites in human and animal subjects.

Table 1. Summary of studies of genistein efficacy in humans

Study Group	Duration	Isoflavone Dosage and Type	Number of subjects per treatment group	Result	References
Postmenopausal women (47-57 years)	12 months	Genistein (54 mg/day)	30	Genistein significantly increased BMD in the femur and lumbar spine relative to placebo.	(Morabito, 2002 #67)
Postmenopausal women (no age available)	12 months	Genistein (54 mg/day)	30	At 1 year, RANKL/OPG ratio was lower in genistein group, which had a negative correlation with 1 year change in femoral neck BMD. The RANKL-OPG system may mediate beneficial effects of genistein on bone remodeling.	(Crisafulli, 2004 #69)
Postmenopausal women (49-67 years old)	24 months	Genistein, (54 mg/day)	198	Mean BMD at femoral neck increased at 1 year and 2 year in genistein treated participants. Twenty-four months of treatment with genistein had positive effects on BMD of lumbar spine and femoral neck in osteopenic postmenopausal women.	(Marini, 2007 #66) (Marini, 2008 #90)
Osteopenic postmenopausal women (49-67 years)	12 and 24 months	Genistein (54 mg/day)	71	Genistein significantly increased BMD in the femoral neck and lumbar spine.	(Atteritano, 2009 #68)

Table 2. Summary of studies of genistein efficacy in animals

Study Group	Duration	Isoflavone Dosage and Type	Equivalent Human Dosage	Number of subjects per treatment group	Result	References
OVX BALB/c mice	1 month	Genistein 0.1, 0.5, 1.5, and 3 mg/day (intra-peritoneal)	--	8	Femur BMDs were completely restored by genistein administration.	(Pie, 2006 #60)
OVX C57BL/6 mice	3 months	Genistein 3 mg/day	--	15	Biomechanical parameters and BMD of genistein group were similar to those seen with estradiol, both of which were significantly superior to control. Cortical bone was only partly improved. Cancellous bone density was significantly increased.	(Sehmisch, 2010 #61)
OVX C57BL/6 mice	1 month	Genistein (25 mg/kg/day)	2.0270 mg/kg/day	6-8	Mice treated with genistein preserved only cross-sectional bone area/cortical surface and periosteal perimeter in cortical bone. Genistein treated mice inhibited trabecular bone loss only partially and did not inhibit bone loss at the cortical level.	(Cano, 2008 #62)
OVX C57BL/6 mice	3 weeks	Genistein 50 mg/kg/day	4.0541 mg/kg/day	9-10	Genistein caused a small but significant increase in BMD at the distal metaphyseal part of the femur in OVX animals.	(Erlandsson, 2005 #56)
OVX female C57BL/6J mice	5 weeks	Genistein 500 mg/kg diet	--	12	Proximal tibial BMD was not significantly induced in response to genistein. Genistein only partially improved the ovariectomy-induced bone structural damages.	(Zhang, 2009 #63)
OVX ddY mice	1 month	Genistein 0.4 mg/day	--	8	BMD of whole femur was higher in genistein group than ovariectomized group alone.	(Wu, 2001 #59)
Male ORX ddY mice	3 weeks	Genistein 0.4-0.8 mg/day	--	6	Genistein prevented orchidectomy-induced BMD decrease in the femur and also prevented ovariectomy-induced loss of trabecular bone volume in femoral distal metaphysis.	(Ishimi, 2002 #65)

Study Group	Duration	Isoflavone Dosage and Type	Equivalent Human Dosage	Number of subjects per treatment group	Result	References
OVX athymic mice	22 weeks	Genistein 10 mg/kg/day	0.8108 mg/kg/day	10-12	Genistein increased femur BMD and femur biomechanical strength.	(Power, 2006 #58)
ORX ddY male mice	3 weeks	Genistein 0.4 mg/day and 0.8 mg/day	--	8	Administration of genistein significantly suppressed post-orchidectomy reduction in distal and whole femur BMD in dose-dependent manner. Genistein at 0.4 mg/day is the minimal dose sufficient for bone-protective effect in ORX mice.	(Wu, 2003 #64)
OVX female Wistar rats	3 months	Genistein 10 microgram/g/ day (10 mg/kg)	1.6216 mg/kg/day	10	Femoral BMD was significantly enhanced, as was the metaphyseal femur. Genistein did not protect the femoral diaphysis. Genistein treatment partially prevented the decrease of cancellous bone area and trabeculae and improved trabecular thickness.	(Mathey, 2007 #45)
OVX female Wistar rats	3 months	10 mg/kg/day (subcutaneous); diet low in isoflavone content enriched with genistein 42 mg/kg/day (IDD)	1.6216 mg/kg/day (subcutaneous); 6.8108 mg/kg/day (IDD)	7	Genistein protected against ovariectomy-induced bone loss in all groups. Subcutaneous or dietary administration of genistein stimulated bone formation but did not decrease bone resorption.	(Hertrampf, 2009 #49)
OVX female Sprague Dawley rats	3 weeks	Genistein 1, 5, 25 microgram/g/ day (1,5,25 mg/kg/day)	0.1622 mg/kg/day, 0.8108 mg/kg/day, 4.0541 mg/kg/day	6	Genistein (5 and 25 microgram/g) protected against ovariectomy-induced loss of BMD in the tibia. 1 microgram/g did not affect BMD loss. Genistein partially prevented post-ovariectomy distal femur cancellous bone loss and trabecular separation increase.	(Fanti, 1998 #46)
OVX female Sprague Dawley rats	1 month	Genistein 12 mg/kg/day	1.9459 mg/kg/day	8	Genistein increased femoral BMD of OVX rats by 5%.	(Nakajima, 2001 #38)

Table 2. (Continued)

Study Group	Duration	Isoflavone Dosage and Type	Equivalent Human Dosage	Number of subjects per treatment group	Result	References
OVX Sprague Dawley rats	15 weeks	Genistein 5 mg/kg/day	0.8108 mg/kg/day	10	Genistein did not prevent ovariectomy-induced deterioration of microstructure or decrease in BMD in vertebral trabecular bone. Genistein preserved biomechanical quality of trabecular bone.	(Dai, 2008 #52)
OVX Sprague Dawley rats	3 months	Genistein 20 mg/kg/day	3.2432 mg/kg/day	10	Genistein significantly increased femoral BMD, trabecular thickness, and trabecular area in proximal tibia. Genistein also significantly increased trabecular number.	(Yang, 2014 #39)
OVX Sprague Dawley rats	3 months	Genistein 4.5 mg/kg/day and 9 mg/kg/day	0.7297 mg/kg/day and 1.4595 mg/kg/day	50	Genistein reduced/recovered ovariectomy-induced BMD loss in femur and tibia at both doses.	(Wang, 2006 #40)
OVX Sprague Dawley rats	3 months	Genistein 4.5, 9, 18 mg/kg/day	0.7297 mg/kg/day, 1.4595 mg/kg/day, 2.9189 mg/kg/day	20	Treatment with all three levels of genistein recovered ovariectomy-induced loss of femur and tibia BMD.	(Miao, 2012 #44)
OVX Sprague Dawley rats	3 months	Genistein 45 mg/kg/day	7.2973 mg/kg/day	10	BMD increased significantly after treatment with genistein.	(Li, 2003 #42)
OVX Sprague Dawley rats	3 months	Genistein 1 and 10 mg/kg/day aglycone (subcutaneous)	0.1622 mg/kg/day and 1.6216 mg/kg/day	12	Genistein (10 mg/kg) showed a greater increase in BMD and BMC than all other treatments.	(Bitto, 2008 #41)
OVX Sprague Dawley rats	1.5 months	Genistein 27, 54, 200, 500, 1000 mg/0.235 kg/day; aglycone form	18.6314 mg/kg/day, 37.2628, mg/kg/day, 138.0104 mg/kg/day, 345.0259 mg/kg/day, 690.0518 mg/kg/day	4	Most effective dose of genistein was 54 mg/day. Higher doses yielded no further bone improving effect.	(Bitto, 2011 #43)

Study Group	Duration	Isoflavone Dosage and Type	Equivalent Human Dosage	Number of subjects per treatment group	Result	References
OVX Sprague Dawley rats	4 days or 5 weeks	Genistein 0.5 mg/kg/day	0.0811 mg/kg/day	--	Genistein showed no bone sparing effects.	(Dodge, 1996 #54)
Sprague Dawley rats	1.5 months	Genistein 10 mg/kg/day and 50 mg/kg/day	1.6216 mg/kg/day and 8.1081 mg/kg/day	10	At both doses, genistein significantly increased BMD and bone volume and resulted in thicker subchondral trabecular bone.	(Li, 2012 #50)
OVX Long-Evans rats	5 months	485 microgram/day and 970 microgram/day	--	7-12	Oral genistein did not affect cancellous and cortical bone mass and bone architecture.	(Turner, 2013 #53)
OVX Female Wistar rats	3 weeks	Genistein 10 mg/kg/day	1.6216 mg/kg/day	6	Genistein significantly increased trabecular BMD (tibia) relative to OVX control group.	(Hertrampf, 2008 #48)
OVX Female Wistar rats	3 months	Genistein 10 micrograms/g/day (10 mg/kg/day)	1.6216 mg/kg/day	13	Genistein prevented diaphyseal BMD loss in femur.	(Picherit, 2000 #129)
OVX female Wistar rats	3 weeks	Genistein 10 mg/kg/day	1.6216 mg/kg/day	5	Genistein treated rats with opportunity of voluntary wheel running had significant reduction in ovariectomy-induced BMD decrease. In absence of physical activity, this effect was not preserved.	(Hertrampf, 2007 #47)
OVX Wistar rats	1 month	Genistein 5 mg/kg/day	0.8108 mg/kg/day	6	Genistein did not improve bone mineralization.	(Sliwinski, 2009 #51)
OVX rats	5 weeks	Genistein 40 mg/kg/day	6.4865 mg/kg/day	12	Genistein maintained vertebral body compression strength levels (96.8%) relative to estrogen treated group.	(Sehmisch, 2010 #3282)
OVX rats	2 weeks	Genistein 0.5 mg/day (low), 1.6 mg/day (intermediate), and 5 mg/day (high) dose	--	--	Lower doses of genistein are approximately equally effective to estrogen in retention of cancellous bone tissue. High doses are less effective.	(Anderson, 1998 #55)

4.1.C. Synergism with Genistein Treatment

Combined treatment of genistein dosage and moderate exercise may have more beneficial effects than independent interventions on prevention of ovariectomy-induced bone loss in rats and mice (Nakajima et al., 2001) (Wu et al., 2001) (Hertrampf et al., 2007) or orchidectomy-induced bone loss in mice (Wu et al., 2003). The combination of zinc and genistein has been found to show a synergistic effect on bone formation (Uchiyama and Yamaguchi, 2007a) (Gao and Yamaguchi, 1998) (Yamaguchi, Gao, and Ma, 2000) (Yamaguchi, 2012) (Karieb and Fox, 2012). Additionally, the combination of genistein and zinc has been shown to augment inhibitory effect of phytoestrogens on osteoclastic cell formation and function (Uchiyama and Yamaguchi, 2007b) (Karieb and Fox, 2012). The combination of genistein and zinc may also have potent stimulatory effects on apoptotic cell death in osteoclasts (Uchiyama and Yamaguchi, 2007b).

4.2. Cell Type Specific Effects

The rat, mouse, and human ER exist as two subtypes, ERα and ERβ. The C-terminal ligand-binding domain and N-terminal transactivation domain of ERα and ERβ differ, which causes differences in binding affinity by phytoestrogens, such as genistein and daidzein (Kuiper et al., 1998). Genistein has potent transactivation activity for both ERα and ERβ, relative to estradiol, in concentrations over 100 nm, but it binds with much higher affinity to ERβ than to ERα (Setchell and Lydeking-Olsen, 2003) (Hwang et al., 2006) (Kuiper et al., 1998).

Therefore, the effects of genistein in stimulating bone formation and inhibiting bone resorption are primarily mediated through the ER, particularly ERβ (Somjen et al., 2011) (Garcia Palacios et al., 2005) (M. H. Liao et al., 2014) (Y. Q. Li et al., 2012) (Ming, Chen, and Xian, 2013) (Salvatori et al., 2009) (Tang, Zhu, Liu, Nicholson, and Ni, 2008) (Rickard et al., 2003) (Kuiper et al., 1998) (Chrzan and Bradford, 2007).

It has been demonstrated that genistein has a biphasic effect on osteogenesis. At low doses (0.5 mg/day in OVX rats, ≤ 1 μM aka 0.1 to 10 μM), genistein tends to act through ER-mediated pathways to stimulate osteogenesis (Anderson et al., 1998) (Z. C. Dang, Audinot, Papapoulos, Boutin, and Lowik, 2003), and at high concentrations (>1 μM), it primarily acts through ER-independent pathways to down regulate osteogenesis and upregulate adipogenesis (Z. C. Dang et al., 2003) (Anderson et al., 1998). Activation of peroxisome proliferator-activated receptor gamma (PPARγ) by high concentrations of genistein down-regulates its estrogenic transcriptional activity, while activation of ERα or ERβ by genistein down-regulates PPARγ transcriptional activity (Z. C. Dang et al., 2003). In a six-week study of genistein (10 and 50 mg/kg), administration to female Sprague Dawley rats, low-dose genistein significantly increased expression of alkaline phosphatase (ALP), osteocalcin (OC), and OPG but decreased the expression of RANKL and RANKL/OPG ratio (Y. Q. Li et al., 2012). In that same study, osteoclasts were prepared from rat mandibular condyle for *in vitro* treatment with genistein (100 nm or 100 μM) for two days. Increased expression of ALP, OC, and OPG but decreased expression of RANKL and RANKL/OPG ratio were observed *in vitro* following the treatment of osteoclasts from the mandibular condyles of the Sprague Dawley rats with the lower dose of genistein (Y. Q. Li et al., 2012). Both *in vitro* and *in vivo*, high dose genistein decreased expression of ALP, OC, OPG, RANKL, and RANKL/OPG ratio. In the rat mandibular condylar subchondral bone, low-dose genistein increased bone formation and inhibited bone resorption while high dose genistein inhibited bone formation and resorption (Y. Q. Li et al., 2012).

In summary, genistein appears to have pro-osteogenic effect acting through ER-dependent pathways at concentrations less than 25 μM and anti-osteogenic effect acting through ER-independent pathways at concentrations greater than 25 μM. Further analysis of the biphasic effect of genistein is required to address the exact point of transition between its osteogenic and antiosteogenic effects. Biphasic effects of genistein are discussed below.

The overall cellular effects of genistein are shown in Figure 1, and Figures 2, 3, 4, 6, and 7 depict its effects at low and high doses in preosteoclasts, mature osteoclasts, preosteoblasts, and mature osteoblasts. It should be noted that the biphasic effect of genistein at low and high concentrations can be seen in the contrast between the pathways activated by genistein in preosteoclasts in Figures 2 and 3, the pathways activated by genistein in preosteoblasts in Figure 6, and the pathways activated by genistein in mature osteoblasts in Figure 7.

4.2. A. Osteoclasts

4.2. A.1. RANKL/RANK Pathway Effects

RANKL expression by osteoblast/stromal cells is stimulated by factors known to stimulate osteoclast formation and activity. RANKL produced by osteoblasts binds RANK on the osteoclast precursor cells in either the soluble or membrane-bound form, causing the recruitment of the adaptor protein TRAF6 (tumor necrosis factor receptor-associated factor 6). This binding activates nuclear factor-κB (NF-κB) and causes it to translocate to the nucleus. NF-κB increases c-Fos expression (Boyce and Xing, 2007).

c-Fos is the initial stimulus for cytoplasmic accumulation of the nuclear factor of activated T cells (NFATc1). NFATc1 is then activated by calcium/calmodulin-dependent signals to translocate to the nucleus and autoamplify its expression.

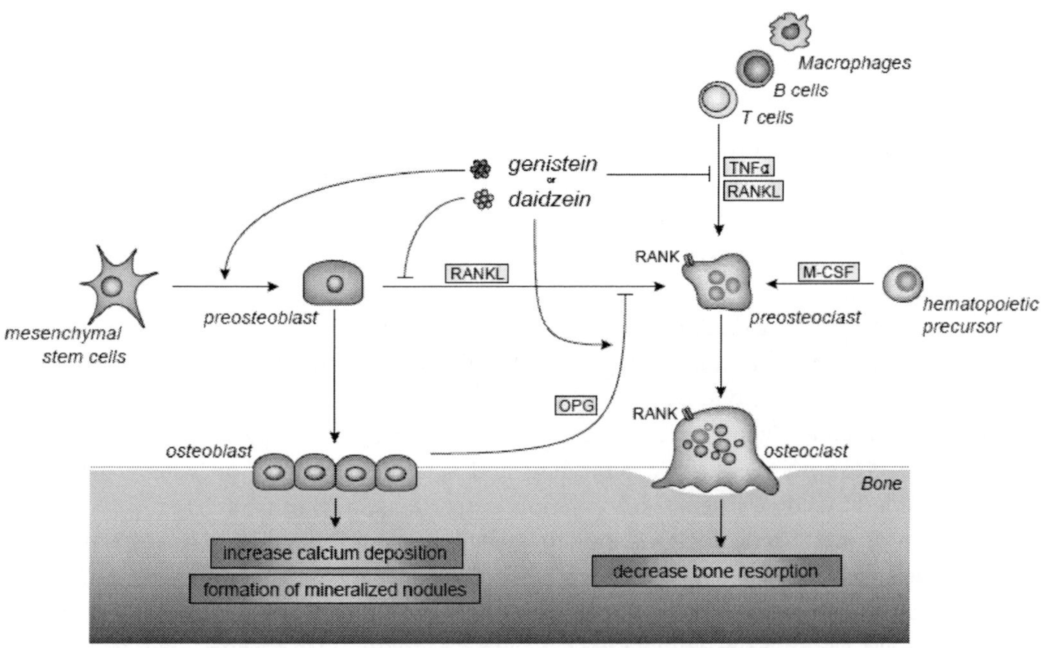

Figure 1. Overview of cellular effects of genistein and daidzein.

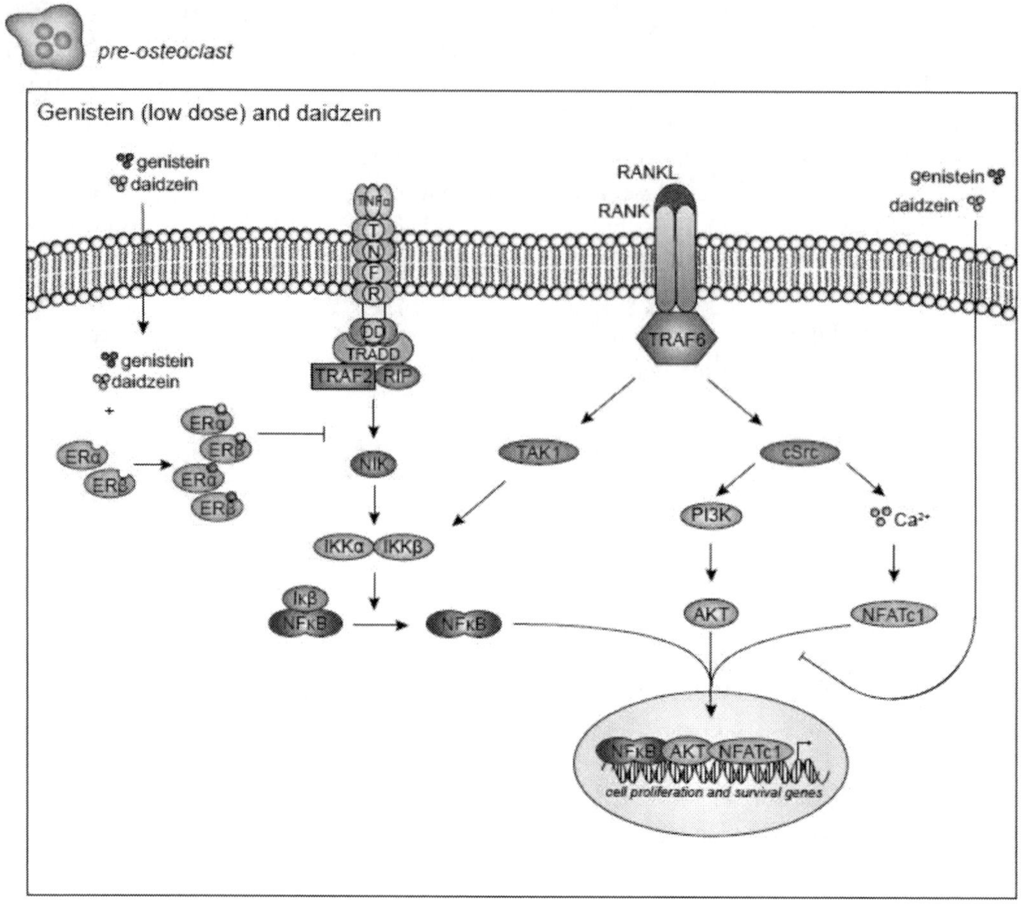

Figure 2. Effects of genistein and daidzein in preosteoclasts at low doses (< 25 μM and < 20 μM, respectively).

This autoamplification of NFATc1 expression is also dependent on c-Fos to stabilize the interaction between NFATc1 and its own promoter (Karieb and Fox, 2011).

NFATc1 requires both of these stimuli to initiate transcription of osteoclastogenic genes and thus begin differentiation of the osteoclast precursor cell into a mature osteoclast (Karieb and Fox, 2011). NFATc1 and c-Fos together regulate the expression of genes such as dendritic cell-specific transmembrane protein (DC-STAMP), which has a key role in later stages of osteoclast differentiation. DC-STAMP enables mononuclear osteoclasts to aggregate and fuse to form large multinuclear osteoclasts.

OPG is a soluble decoy receptor expressed by osteoblasts which binds RANKL on osteoblasts to prevent RANKL from binding and activating RANK on osteoclasts. When RANKL is bound by OPG, osteoclast formation, activation, and survival ceases.

Inhibition of the RANKL/RANK signaling pathway and reduction of the RANKL:OPG ratio may be major mechanisms by which genistein inhibits osteoclast maturation and thus help to explain the positive effects of genistein on BMD (Marini et al., 2008). While direct inhibitory effects of soybean extracts on osteoclast formation are feasible, it is more likely that inhibitory effects on osteoclast differentiation are mediated indirectly through paracrine factors produced by osteoblasts, such as OPG and RANKL (K. Park et al., 2014).

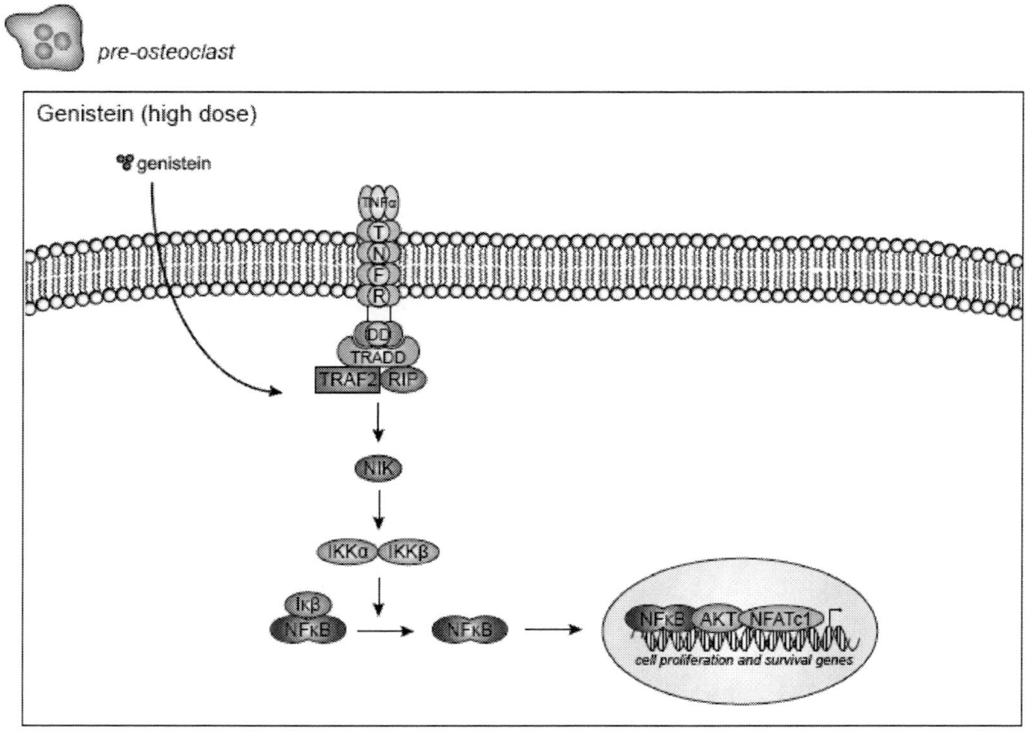

Figure 3. Effects of genistein in preosteoclasts at high doses (≥ 25 μM).

The primary effect of genistein in decreasing osteoclast-like cell number may be mediated through its effects on the RANKL pathway. This has been demonstrated in a study by Uchiyama et al. in which culture with genistein for 24 or 48 hours in the absence of M-CSF and RANKL did not cause significant decrease in osteoclast-like cells; however, culture with genistein in the presence of M-CSF and RANKL did cause significant decrease (Uchiyama and Yamaguchi, 2007b). Genistein (3 μM) has been shown to significantly inhibit NF-κB nuclear localization, part of the RANKL/RANK pathway, in a murine monocytic cell line and also to decrease NF-κB levels in an ER-independent manner (Garcia Palacios et al., 2005) (Ming et al., 2013) (Karieb and Fox, 2011).

Additionally, genistein (100 nm) has been shown to significantly reduce TNF-α induced c-Fos and NFATc1 expression in an ER-dependent manner and to inhibit NFATc1 nuclear translocation (Karieb and Fox, 2011) (Tadaishi, Nishide, Tousen, Kruger, and Ishimi, 2014). Genistein reduces osteoclast formation and activity and has therefore been shown to be mediated via an action on osteoclast differentiation (Karieb and Fox, 2011). These effects of low doses of genistein on preosteoclasts are depicted in Figure 2.

Interestingly, genistein significantly down-regulated the expression and secretion of RANKL at a variety of concentrations *in vitro*: 50 μM (Y. Li et al., 2006) (Yamagishi, Otsuka, and Hagiwara, 2001), 100 nm (Y. Q. Li et al., 2012), 100 μM (Y. Q. Li et al., 2012), and 1 μM to 100 nm (W. F. Chen and Wong, 2006). It also down-regulated the expression and secretion of RANKL in OVX mice (Y. Li et al., 2006) (Zhang et al., 2009) and OVX rats (Y. Q. Li et al., 2012) (Bitto et al., 2008). Genistein has been shown to suppress bone resorption by enhancing the production of OPG by osteoblasts. Studies conducted with OPG-deficient mice demonstrated the effects of genistein on bone resorption *in vivo* were lost completely, suggesting that the effect of genistein is dependent on OPG (J. Liu et al., 2008). Genistein has been reported to significantly induce OPG expression at a variety of concentrations *in vitro*: 50 μM (Y. Li et al., 2006), 100 nm (Y. Q. Li et al., 2012), 30 μM (Yamagishi et al., 2001), 10 nm or 1 μM (Heim et al., 2004), 10 nm (X. Chen, Garner, Quarles, and Anderson, 2003), 10 nm to 1 μM (W. F. Chen and Wong, 2006), 100 nm (Viereck et al., 2002), 10 nm to 1 μM (X. W. Chen, Garner, and Anderson, 2002). Genistein (10 nm to 1 μM) has been demonstrated to induce OPG expression in human osteoblasts in an ER-dependent manner (W. F. Chen and Wong, 2006) (Viereck et al., 2002) (X. W. Chen et al., 2002). Human osteoblast cells (hOB) contain ERα protein, but ERβ protein could not be detected, suggesting that genistein is capable of increasing OPG mRNA and protein levels through ERα (Viereck et al., 2002). Genistein was also shown to enhance OPG expression in OVX mice (Y. Li et al., 2006) and OVX rats (Bitto et al., 2008) (Y. Q. Li et al., 2012). The RANKL:OPG gene expression ratio is indicative of the state of bone remodeling. In states of primarily bone resorption, RANKL:OPG will be increased as pro-osteoclastogenic factors such as RANKL are being produced in greater amounts than anti-osteoclastogenic factors such as OPG, whereas the RANKL:OPG ratio is decreased when osteoblasts shift to inhibit bone resorption and stimulate bone formation. Genistein significantly reduced RANKL:OPG gene expression ratio at a variety of concentrations *in vitro*: 50 μM (Y. Li et al., 2006), 100 nm (Karieb and Fox, 2012), 100 nm (Y. Q. Li et al., 2012), 100 μM (Y. Q. Li et al., 2012), 10 nm or 1 μM (Heim et al., 2004), 100 nm to 1 μM (W. F. Chen and Wong, 2006), 1 nm to 100 nm with maximum effect at 10 nm (Sliwinski et al., 2009). Genistein has also been demonstrated to reduce the RANKL:OPG expression ratio in OVX mice (Y. Li et al., 2006) (Zhang et al., 2009) and OVX rats (Y. Q. Li et al., 2012).

The effect of genistein on decreasing RANKL:OPG gene expression ratio has also been observed in postmenopausal women. In a study of 389 postmenopausal osteopenic women treated with 54 mg/day genistein for 24 months, genistein lowered soluble RANKL level and increased OPG at one and two years, relative to placebo. At the end of the 24 months, genistein produced significant decreases in soluble RANKL:OPG ratio compared to placebo (Marini et al., 2008). In another study, early postmenopausal women were treated with genistein (54 mg/day) for 1 year, which resulted in a lower soluble RANKL:OPG ratio (Crisafulli et al., 2004). This decrease in sRANKL:OPG ratio positively correlated with the 1-year change in urinary deoxypyridinoline and negatively correlated with 1-year change in femoral neck BMD. These results suggest that the sRANKL-OPG system mediates some of the beneficial effects that genistein has on bone remodeling in postmenopausal women.

An overview of the cellular effects of genistein on RANK, RANKL, and OPG is depicted in Figure 1.

4.2.A.2. Immunomodulatory Effects Relevant to the Prevention of Bone Loss

Genistein may also prevent bone resorption by reducing ovariectomy-induced B-lymphopoiesis. Genistein delivered at 0.5 mg/day restored ovariectomy-induced stimulation of bone marrow hemopoiesis to the same level as sham mice (Ishimi et al., 1999). Genistein did not have any effect on B-lymphopoiesis in mice that had not been OVX. In male ORX mice treated with genistein delivered at 0.4-0.8 mg/day for 3 weeks, genistein abolished the orchidectomy-induced increase in pre-B cells (Ishimi et al., 2002). Genistein also has been demonstrated to have inhibitory effects on T cells. Genistein (100 nm) suppressed the maturation of osteoclasts induced by T cell-derived TNF-α (Karieb and Fox, 2013).

Genistein has not been shown to affect T cell number but affected T cell differentiation, reducing expression of TNF-α, IL-1β, IL-6 and RANKL (Karieb and Fox, 2013). This prevention of the pro-osteoclastic action of T cells suggests that genistein may have a role in controlling inflammatory bone loss (Karieb and Fox, 2013).

Furthermore, TNF-α has been shown to promote osteoclast differentiation by increasing RANKL-induced osteoclastogenesis, directly stimulating osteoclast maturation from monocytes, and activating osteoclast formation independent of RANKL signaling (Karieb and Fox, 2011). RANKL activates TRAF2, TRAF3, and TRAF6 while TNF-α predominantly signals via TRAF2 and TRAF3 (Karieb and Fox, 2011). Part of the anti-osteoclastogenic effect of genistein may be mediated through its ability to decrease the release of proinflammatory cytokines such as TNF-α (Comalada et al., 2006) (Huang et al., 2005) (Fanti et al., 1998) (B. Li and Yu, 2003), IL-1β (B. Li and Yu, 2003), and IL-6 (X. W. Chen et al., 2002) (X. Chen et al., 2003) (Rickard et al., 2003) (Suh et al., 2003). Although genistein (10 nm to 1 μM) has been shown to decrease release of IL-6 through an ER-dependent pathway, it remains to be determined whether the decrease in levels of TNF-α and IL-1β are mediated through ER-dependent pathways (X. W. Chen et al., 2002) (X. Chen et al., 2003) (Rickard et al., 2003). Direct inhibition of TNF-α production from monocytes by soy isoflavones may be involved in early decrease of TNF-α levels (Huang et al., 2005). Pie et al. demonstrated that genistein treatment resulted in down regulation of TNF-α, IL-1β, and IL-6 (Pie et al., 2006). Interestingly, lower concentrations of isoflavones (10 nm and 100 nm) were more effective in inhibiting TNF-α levels than higher concentrations (1 μM for isoflavones), which supports a biphasic effect of isoflavone treatment (Huang et al., 2005). The effects of genistein at low and high doses in preosteoclasts are depicted in Figures 2 and 3, respectively.

4.2.A.3. Effects on Intracellular Calcium

Calcium has been shown to play a role in osteoclast differentiation, bone resorption, and gene transcription (Kajiya, 2012). RANKL signaling has been shown to induce oscillations in intranuclear calcium concentrations, leading to in calcium/calcineurin-dependent dephosphorylation and activation of NFATc1, which initiates osteoclast differentiation. An increase in RANKL-induced intracellular calcium has been shown to inhibit cell motility and cytoskeleton resorption in mature osteoclasts, resulting in the suppression of bone resorption. Thus, while calcium signaling activates differentiation in osteoclast precursors, it suppresses bone resorption in mature osteoclasts (Kajiya, 2012). These results suggest that calcium signaling may activate differing pathways in osteoclast precursors and mature osteoclasts, which mediate resorptive and antiresorptive effects, respectively.

One of the anti-osteoclastogenic mechanisms of genistein involves calcium signaling.

Genistein (50 μM and 100 μM with 100 μM yielding greater effect) has been shown to reversibly elevate intracellular calcium in rat osteoclasts via activation of a dihydropyridine (DHP) insensitive, nonspecific calcium entry pathway (Kajiya, Okabe, Okamoto, Tsuzuki, and Soeda, 2000). This genistein-mediated increase in intracellular calcium has been demonstrated to inhibit resorption by rapidly disrupting the actin ring formation that serves as a marker for the osteoclast resorbing state (Kajiya et al., 2000).

Because genistein had effects on the actin cytoskeleton, it is likely to also exert its effect during later osteoclast differentiation or in mature osteoclasts. Another study by Gao and colleagues demonstrated that genistein might act on late stage pre-osteoclast differentiation and/or mature osteoclasts (Gao and Yamaguchi, 1999b).

In this study, genistein (1 μM or 10 μM) markedly weakened the increase in osteoclasts in rat marrow cells cultured with vehicle, PTH, or prostaglandin E2 (PGE2) and caused a significant decrease in the number of osteoclasts. The suppressive effect of genistein was completely abolished by the presence of calcium-dependent protein kinase inhibitors.

These findings indicate that genistein exerts its suppressive effect on osteoclasts via an intracellular calcium signaling pathway, whereas tamoxifen was unable to inhibit the effect of genistein, indicating that this effect is not ER-dependent (Gao and Yamaguchi, 1999b). These effects of genistein on intracellular calcium in mature osteoclasts are shown in Figure 4.

Another ER-independent pathway which may cause membrane depolarization, an increase in intracellular calcium, and inhibition of osteoclastic bone resorption was described by Okamoto et al. (Okamoto, Okabe, and Kajiya, 2001).

In this study, concentrations of genistein greater than 1 μM were shown to significantly inhibit the inward rectifier potassium current (IKir) in rat osteoclasts. Removal of genistein partially restored the current. Inhibition of IKir by genistein does not depend on protein tyrosine kinase inhibition, intracellular calcium changes, or secondary interaction with protein kinase A (PKA) or protein kinase C (PKC). It is inconsistent with the time course of responses mediated by intracellular ER, which take at least 30 minutes to be activated.

Therefore, it has been suggested that the inward rectifier K+ channels in osteoclasts may play a role in maintaining proton (H^+) transport to the surface of bone opposite the ruffled border membrane, a function necessary for resorption. The effects of genistein on the IKir are shown in Figure 4.

4.2.A.4. Effects on Cell Proliferation and Apoptosis

Genistein has been demonstrated to significantly decrease the number of osteoclast-like cells and to inhibit formation of osteoclast-like cells at a variety of concentrations ranging from 1 nm to 50 μM (Uchiyama and Yamaguchi, 2007b) (Yamagishi, 2001 #72) (Comalada, 2006 #65) (Hwang, 2006 #61) (Lee, Kim, and Jang, 2014) (Sliwinski et al., 2009) (Karieb and Fox, 2011) (Gao and Yamaguchi, 1999b) (Sliwinski, Folwarczna, Janiec, Grynkiewicz, and Kuzyk, 2005) (Tadaishi et al., 2014) (B. Li and Yu, 2003).

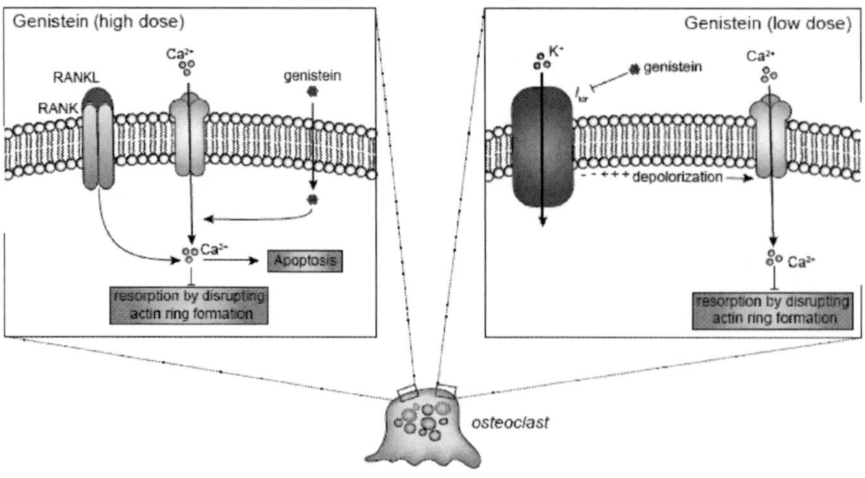

Figure 4. Effects of genistein in mature osteoclasts at low and high doses (>1 μM and >50 μM, respectively).

Interestingly, in the study by Sliwinski et al., genistein was shown to decrease the number of osteoclasts at 10 nm due to decreasing their viability and at 10 μM due to attenuation of formation of osteoclasts (Sliwinski et al., 2005). Hwang et al. and Karieb et al. demonstrated that inhibitory effects of genistein on osteoclast formation may be ER-dependent (Hwang, 2006 #61) (Karieb and Fox, 2011).

It has been reported that at concentrations greater than 10 μM, genistein increases cell death in osteoclast-like cells (Garcia Palacios et al., 2005) (Garcia Palacios et al., 2005) (Garcia Palacios et al., 2005) (Tadaishi et al., 2014), (Garcia Palacios, 2005 #68), (Hwang, 2006 #61). Additionally, Hwang et al. demonstrated that this effect may be ER-mediated.

4.2. A.5. Effects on Oxidative Stress

Age-related decline in bone mass is associated with an increase in oxidative stress and an increase in osteoclastic activity, relative to osteoblastic activity, such that intracellular ROS level increases during RANKL-induced osteoclast differentiation. ROS is endogenously produced either by nicotinamide adenine dinucleotide phosphate (NADPH) oxidase (Nox1) or as byproducts of the mitochondrial electron transport train during the differentiation of osteoclast precursors into mature osteoclasts, and the ROS generated is removed by antioxidant enzymes such as superoxide dismutase (SOD) and heme oxygenase-1 (HO-1).

Suppression of the ROS production by downregulation of Nox1, decreased disruption of the mitochondrial electron transport chain, or by upregulation of antioxidant enzymes may suppress osteoclast differentiation (Lee et al., 2014).

Genistein (20 μM) has been demonstrated to decrease the levels of ROS, confirming the cellular antioxidant capacity, anti-osteoclastogenic, and anti-osteoporotic effects of genistein (Lee et al., 2014).

In a study by Lee et al., genistein suppressed translation and activation of Nox1 and disruption of mitochondrial electron transport chain system while it enhanced induction of phase II antioxidant enzymes, such as SOD1 and HO-1 at concentrations of 1-10 μM (Lee et al., 2014). Therefore, the effects of genistein to decrease ROS levels may suppress osteoclast differentiation and on a larger scale, may translate to decreased oxidative stress and reduction of aging-related bone loss.

4.2. B. Osteoblasts

Genistein has been demonstrated to stimulate osteogenic differentiation and maturation of bone marrow stromal progenitor cells (BMSCs) and osteoblasts. Its effects in these cells are mediated through activation of ER, p38 mitogen-activated protein kinase (MAPK)-Runx2, and nitric oxide/cyclic guanosine monophosphate (NO/cGMP) pathways (Ming et al., 2013).

The effects of genistein on BMSCs undergoing osteogenic commitment are outlined below, and Figures 6 and 7 demonstrate the effects of genistein in preosteoblasts and mature osteoblasts at varying doses.

4.2.B.1. Effects on Proliferation and Commitment

This stage of osteogenesis is accompanied by increase in cell number and inverse regulation of Cbfa-1 and PPARγ (Heim et al., 2004). Core binding factor 1 (Cbfa1)/Runx2, belonging to the runt domain transcription factor family, is a master regulator of the early stages of osteoblastic differentiation.

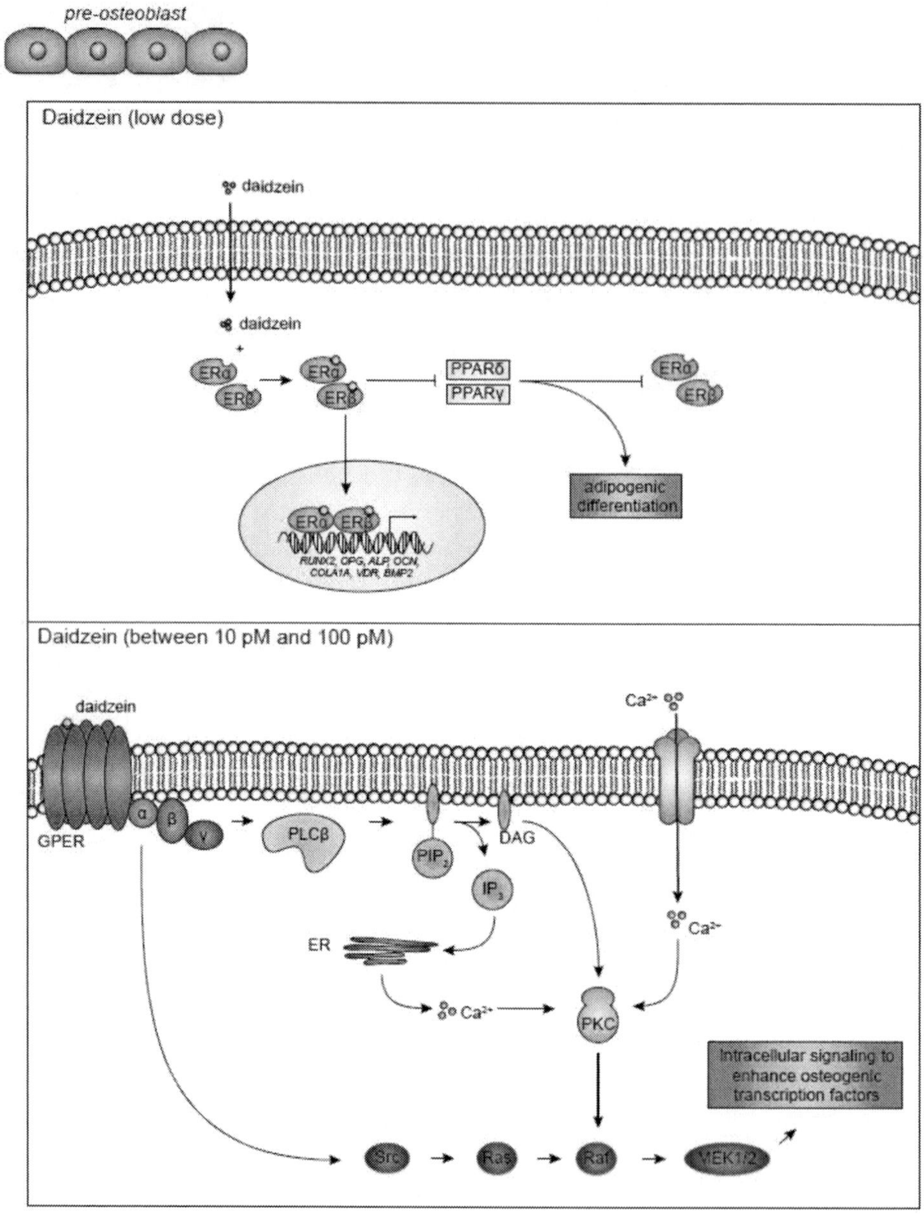

Figure 5. Effects of daidzein in preosteoblasts at low doses (less than 20 μM).

Runx2 expression is increased during early stages of osteoblast differentiation when it promotes formation of immature osteoblasts from MSCs, and Runx2 expression is down-regulated during formation of mature osteoblasts (Fakhry, Hamade, Badran, Buchet, and Magne, 2013). In mice, Runx2 deficiency leads to the formation of skeleton lacking osteoblasts (Fakhry et al., 2013). Cbfa/Runx2, in turn, regulates alkaline phosphatase (ALP), bone sialoprotein (BSP), osteocalcin (OC), and osteopontin (OPN) expression.

With respect to genistein, mRNA expression of Runx2 has been shown to be increased by concentrations of genistein between 1 μM and 10 nm (Z. C. Dang et al., 2003) (Heim et al., 2004) (Q. C. Liao, Xiao, Qin, and Zhou, 2007) (Pan et al., 2005) (J. Dai et al., 2013).

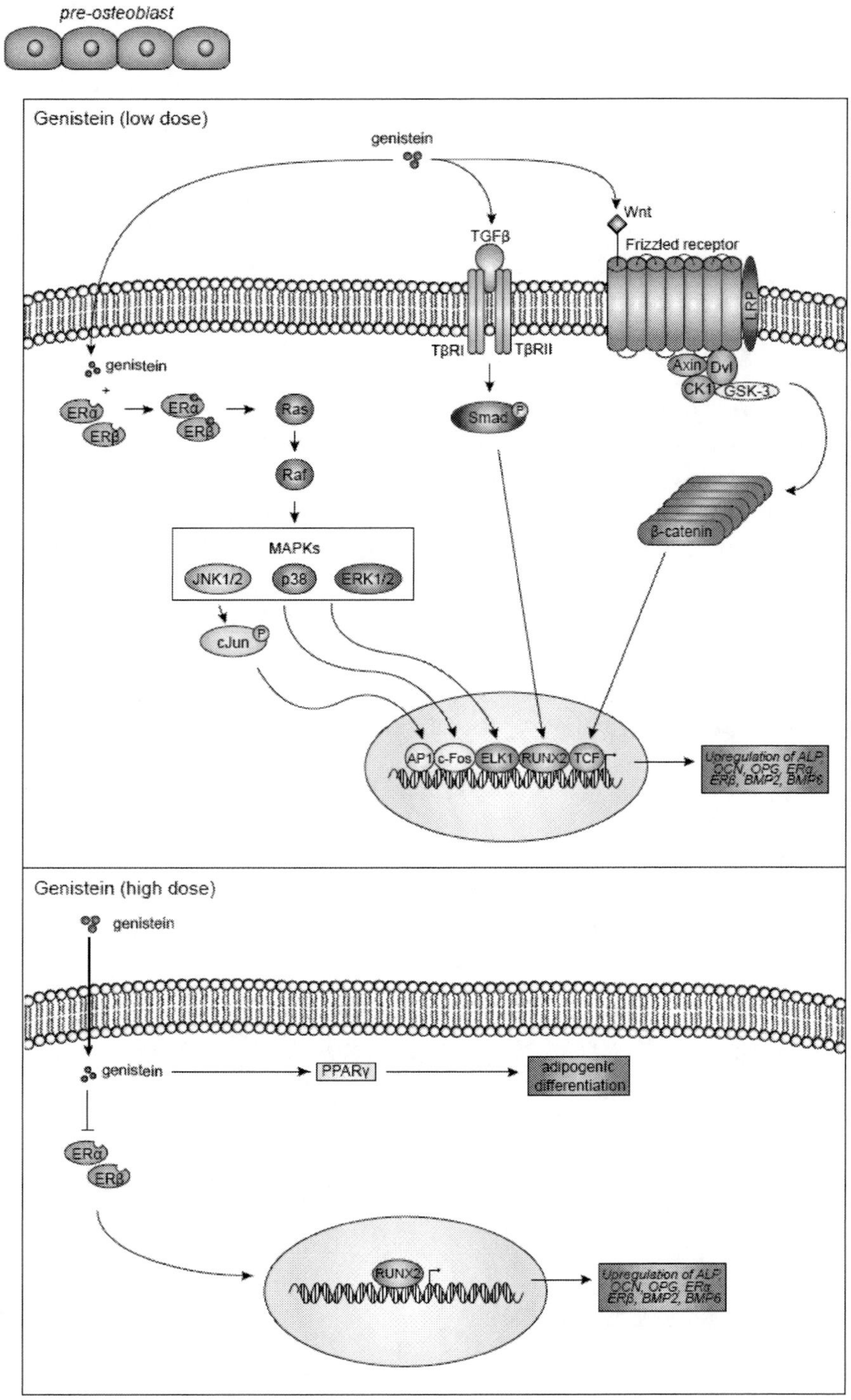

Figure 6. Effects of genistein in preosteoblasts at low and high doses (<25 μM and ≥25 μM, respectively).

Genistein was also shown to enhance Cbfa1 DNA binding activity (Q. C. Liao et al., 2007). In contrast, genistein concentrations of 25 μM decreased Runx2 expression (Z. C. Dang et al., 2003).

Genistein has not been demonstrated to have a consistent effect on osteoblast proliferation. While some studies have shown that genistein caused an increase in BMSC and osteoblast number/proliferation (Karieb and Fox, 2012) (Suh et al., 2003) (Sugimoto and Yamaguchi, 2000) (Okumura, Yoshikawa, Iida, Nonomura, and Takakura, 2006) (Pan et al., 2005) (Somjen et al., 2011) (J. Dai et al., 2013), many studies have also shown that genistein has no effect on osteoblast proliferation (Kanno, Hirano, and Kayama, 2004) (Rickard et al., 2003) (W. F. Chen and Wong, 2006) (Tang, Zhu, Liu, Wang, and Ni, 2011) or that it inhibits proliferation (Salvatori et al., 2009) (Rickard et al., 2003) (Morris, Thorpe, Ambrosio, and Santin, 2006). Therefore, additional studies are necessary to determine definitively the effect of genistein on osteoblast proliferation.

4.2. B.2. Effects on MSC Differentiation

This stage of osteogenesis involves upregulation of ALP expression and collagen type I. Genistein has been shown to stimulate ALP activity at a variety of concentrations, ranging from 10 nm to 10 μM with the majority of studies supporting stimulation of ALP activity at 1 μM (Z. C. Dang et al., 2003) (Sugimoto and Yamaguchi, 2000) (Heim et al., 2004) (Kanno et al., 2004) (Okumura et al., 2006) (Pan et al., 2005) (Q. C. Liao et al., 2007) (Rickard et al., 2003) (W. F. Chen and Wong, 2006) (J. Dai et al., 2013) (Gao and Yamaguchi, 1999a) (Yamaguchi et al., 2000) (J. Dai et al., 2013) (Y. Q. Li et al., 2012). Genistein treatment has also been demonstrated to increase ALP levels in OVX rats (B. Li and Yu, 2003) (Y. Q. Li et al., 2012) (Bitto et al., 2008) and postmenopausal women (Morabito et al., 2002) (Marini et al., 2007).

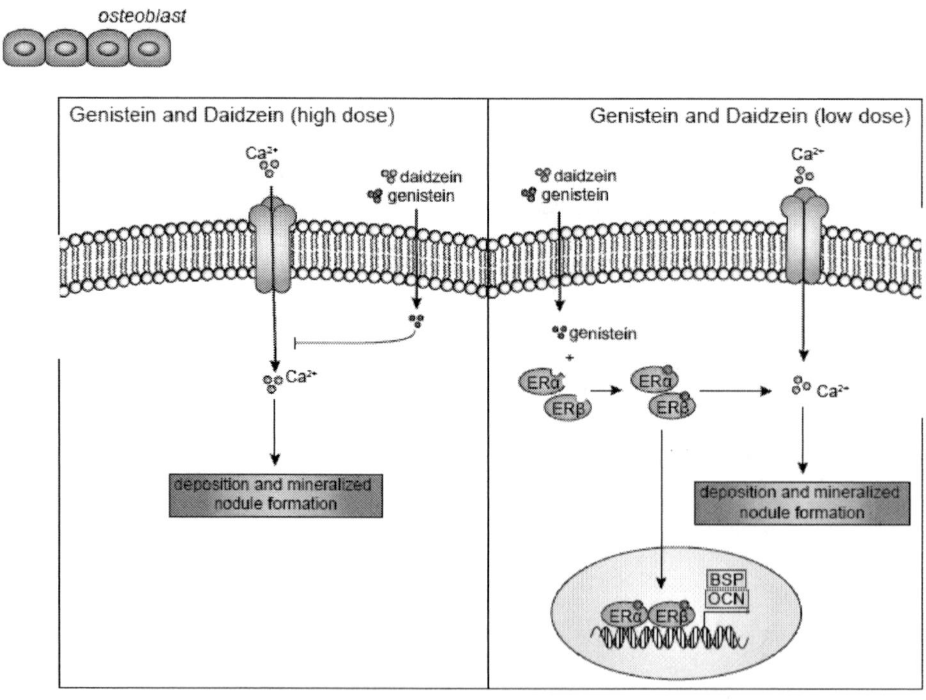

Figure 7. Effects of genistein and daidzein in mature osteoblasts at low and high doses (low dose being <25 μM and <20 μM, respectively, and high dose being ≥25 μM and ≥20 μM, respectively).

It has been demonstrated that ALP activity induced by genistein is mediated through an ER-dependent pathway. Chen et al. showed that genistein induced ALP in an ER-dependent post-transcriptional manner (W. F. Chen and Wong, 2006). ALP expression was increased significantly by genistein in an ERβ expressing mature osteoblastic cell line but not in an ERα expressing mature osteoblastic cell line. These results indicate that an increase in ALP may be mediated specifically through ERβ (Salvatori et al., 2009).

Genistein has been demonstrated to enhance ALP activity between 10 nm and 10 μM, whereas at concentrations of 25 μM or higher genistein has been shown to inhibit ALP activity (Z. C. Dang et al., 2003). The study by Li et al. supported this biphasic effect proposed by Dang et al. and demonstrated that 100 μM (high dose) genistein decreased ALP expression (Y. Q. Li et al., 2012). Genistein was demonstrated to have a stimulatory effect on collagen expression, which is necessary for bone matrix formation. Genistein (100 nm) significantly enhanced type I collagen mRNA expression (Karieb and Fox, 2012). In another study, osteoblasts synthesized high levels of collagen when treated with concentrations of genistein ranging from 5 μM to 30 μM (Morris et al., 2006).

4.2. B.3. Effects on Maturation

The maturation stage of osteogenesis is characterized by increasing levels of OP, BSP-2 and induction of mineralized nodule formation (Heim et al., 2004).

It has been demonstrated the genistein enhances nodule formation and calcium deposition at a variety of concentrations of genistein ranging from 10 nm to 10 μM with the majority of studies showing enhanced nodule formation and calcium deposition at 1 μM and 10 μM (Z. C. Dang et al., 2003) (Karieb and Fox, 2012) (Okumura et al., 2006) (Pan et al., 2005) (Q. C. Liao et al., 2007) (Morris et al., 2006) (Chang, Jin, Jin, Gu, and Zhou, 2003) (Gao and Yamaguchi, 1999a) (Yamaguchi and Gao, 1998) (Yamaguchi and Weitzmann, 2009) (Strong et al., 2014). The effect of genistein to enhance osteoblast mineralization has been shown to be prevented by ICI 182780, indicating that genistein directly enhances osteoblastic mineralization by an ER-dependent mechanism (Karieb and Fox, 2012) (Chang et al., 2003). While genistein enhanced nodule formation and calcium deposition at concentrations less than 25 μM, genistein has been shown to inhibit nodule formation and calcium deposition at concentrations of 25 μM or higher (Z. C. Dang et al., 2003).

Bone matrix is made up of inorganic hydroxyapatite and organic constituents, such as type I collagen, and a wide range of noncollagenous proteins including osteocalcin (Karieb and Fox, 2012). Osteocalcin is a protein unique to bone that has been shown to be involved in mineralization and is used as a sensitive marker for bone activity and mature osteoblast differentiation. Osteocalcin is a downstream gene of Runx2/Cbfa1 (Pan et al., 2005) and has been shown to be increased by genistein at concentrations between 10 nm and 10 μM with the majority of studies demonstrating increased osteocalcin at 1 μM genistein (Pan et al., 2005) (Karieb and Fox, 2012) (Z. C. Dang et al., 2003) (Y. Q. Li et al., 2012) (Chang et al., 2003) (J. Dai et al., 2013) (M. H. Liao et al., 2014) (Salvatori et al., 2009). The potential for genistein to increase osteocalcin levels has been demonstrated to be ER-mediated (Chang, 2003 #88) (M. H. Liao et al., 2014). Genistein has also been shown to increase osteocalcin levels in OVX rats, OVX mice, and postmenopausal women at low dosages (B. Li and Yu, 2003) (Y. Q. Li et al., 2012) (Pie et al., 2006) (Morabito et al., 2002). In contrast, high dosages of genistein inhibited osteocalcin expression. Mice treated with 1.5 or 3 mg/day of genistein (higher concentrations) showed lower serum levels of osteocalcin (Pie et al., 2006).

High-dose genistein (50 mg/kg and 100 μM) decreased expression of osteocalcin (Y. Q. Li et al., 2012). mRNA expression of osteocalcin was decreased by 25 μM genistein administration (Z. C. Dang et al., 2003).

4.2.C. Adipose-Derived Stromal Cells (ASCs)

Mesenchymal stem cells in the bone marrow have the potential to differentiate into multiple types of cells. Two such cell types are adipose-derived stem cells (ASCs) and bone marrow-derived stem cells (BMSCs) which are both capable of differentiating through the osteoblastic lineage. A study conducted by Strong et al. demonstrated that ASCs treated with genistein underwent a 1.4 fold increase in osteogenic differentiation relative to vehicle-treated ASCs (Strong et al., 2014). The Strong et al. study indicates that compounds such as genistein may target not only BMSCs but also ASCs to increase bone formation.

4.3. Molecular Effects

The effects of genistein are mediated through a variety of molecular pathways. One such pathway is the epidermal growth factor receptor (EGFR) pathway, which has been shown to induce proliferation in human BMSCs and may play a role in inhibition of differentiation of BMSCs into the osteoblastic lineage (Matsuda, Yokoyama, Takeshita, and Watanabe, 1998) (Chien, Lin, and Cho, 2000) (Antosz, Bellows, and Aubin, 1989) (Tamama, Fan, Griffith, Blair, and Wells, 2006). Approximately 1 μM genistein treatment of osteosarcoma cells stably expressing ERα or ERβ led to reduced proliferation and cell cycle progression and enhanced apoptosis in both cell lines. Genistein modified the transcriptional activity of the EGFR gene promoter by binding both ERs in ligand and receptor specific way. Through ER-binding, genistein inhibited EGFR mRNA expression transiently. Genistein also progressively reduced EGFR protein level, until it was almost undetectable. EGFR inhibition reduced cell growth by cell cycle arrest, activation of apoptosis, and stimulation of cell differentiation (Salvatori et al., 2009).

While the effect of genistein on proliferation of osteoblasts remains to be definitively determined, a study by Pan et al. has shown that genistein stimulates proliferation and osteoblastic differentiation of mouse BMSCs at least in part through activation of the ER-dependent NO/cGMP pathway (Pan et al., 2005). cGMP, through G-kinase activation, modulates c-Fos expression, which is a part of the activator protein-1 (AP-1) transcription factor complex. The AP-1 class of transcription factors is composed of heterodimers of Fos-related factors such as c-Fos and Jun proteins, which regulate osteoblast differentiation through targeting the OC, BSP, COL1A2, and ALP gene promoters, at which AP-1 physically and functionally interacts with Runx2 to regulate gene expression (Wagner, 2010) (Jensen, Gopalakrishnan, and Westendorf, 2010). (Jensen et al., 2010) (Bozec et al., 2010) (Bozec et al., 2013). In contrast to the anabolic effects of low doses, genistein at higher doses (10 μM) was not as efficient (Pan et al., 2005) (Anderson et al., 1998). Genistein may also act through estrogen independent pathways, since its effect on BMSC proliferation was not completely blocked with ER antagonists (Pan et al., 2005).

Another important signaling pathway that has been shown to be activated by genistein is the MAPK pathway. Studies have demonstrated that 10 μM genistein can trigger phosphorylation of MAPK in osteoblastic cells (M. H. Liao et al., 2014).

MAPKs are serine-threonine specific protein kinases containing three enzymes: extracellular signal-regulated kinase 1/2 (ERK 1/2), p38 mitogen-activated protein kinase (MAPK) and c-Jun N-terminal kinase 1/2 (JNK 1/2) (M. H. Liao et al., 2014). After phosphorylation, ERK1/2 and p38 MAPK activate NF-κB, resulting in the translocation of NF-κB into the nucleus and transcription of osteoblastic genes.

In contrast, activated JNK 1/2 can phosphorylate c-Jun, which can bind to c-Fos to form heterodimeric AP-1. NF-κB and AP-1 are two transcription factors that regulate osteoblast differentiation and mineralization (Kim, Lee, Kim, Min, and Kim, 2008) (Tan et al., 2012) (M. H. Liao et al., 2014). Through the activation of the MAPK/NF-κB/AP-1 axis, genistein can also induce ERα gene expression (M. H. Liao et al., 2014). These effects of genistein to activate MAPK are shown in Figure 6.

In a study by Liao et al., genistein (10 nm to 1 μM) dose-dependently led to rapid and sustained activation of the p38 MAPK pathway in BMSCs. Genistein has been shown to be acting at least partly through this pathway to induce osteoblast maturation and Runx2 activation as treatment with a p38 MAPK inhibitor diminished these effects in mouse BMSC cultures (Q. C. Liao et al., 2007).

This function of genistein is particularly relevant to its role in bone formation as the p38 MAPK pathway has been shown to be essential for normal bone formation in mice (Greenblatt, 2010). In the Liao et al. study, genistein did not affect ERK activity, another important member of the MAPK pathway (Q. C. Liao et al., 2007).

The pro-osteogenic effects of genistein may be mediated through NF-κB activity in preosteoblasts. It has been shown that genistein (1 or 10 μM) significantly increased NF-κB activity in preosteoblastic cells (Yamaguchi and Weitzmann, 2009) and increased the translocation of NF-κB and c-Jun from cytoplasm to nucleus (Q. C. Liao et al., 2007).

Genistein promotion of NF-κB induction is suggested to be through an indirect mechanism involving transcriptional regulation of a factor that induces NF-κB. These pathways ultimately regulate osteoblast differentiation and differentiation suggesting that genistein acts through NF-κB induction to at least partly mediate its anabolic effects on bone.

Bone morphogenetic proteins (BMPs) are potent inducers of osteoblastic differentiation (Fakhry et al., 2013). In the study by Dai et al., genistein was shown to increase human BMSC expression of bone morphogenetic proteins including BMP1, BMP2, BMP3, BMP4, BMP5, BMP6, and BMP7 (J. Dai et al., 2013).

In another study investigating the effect of genistein on gene expression in OVX mice, the upregulation of BMP4 was noted (Pie et al., 2006). BMPs require the presence of extracellular matrix collagen to enhance osteoblastic differentiation and osteogenic gene expression, and genistein has also been shown to upregulate genes for collagen production (COL15A1, COL2A1, COL16A1, COL1A1, COL14A1) (Dai, 2013 #27) (M. H. Liao et al., 2014). BMP-dependent SMAD signaling to induce Runx2/Cbfa1 expression and BMP-induced collagen synthesis to form an extracellular matrix for bone mineralization have been shown to be two major mechanisms by which genistein induces osteoblast-specific gene expression and osteogenesis (J. Dai et al., 2013).

The effect of genistein on BMP2 expression has been well studied. BMP2, BMP6, and BMP7 are commonly known as osteogenic BMPs based on their ability to ectopically induce bone formation, and loss of BMP2 in mice has been shown to produce a phenotype similar to osteoporosis (Bandyopadhyay, Yadav, and Prashar, 2013).

In genistein treated OVX rats, tibial trabecular BMD was significantly higher than that of vehicle treated OVX rats. These observations correlated positively with BMP2 protein expression in the tibia (Hertrampf et al., 2007). Genistein treated animals demonstrated the highest BMP-2 expression (Hertrampf et al., 2007).

It has also been reported that BMP2 (J. Dai et al., 2013) and BMP6 (J. Dai et al., 2013) (M. H. Liao et al., 2014) expression was significantly increased by an ERα-dependent mechanism following *in vitro* and/or *in vivo* genistein administration. Genistein has also been shown to interact with parathyroid hormone/parathyroid hormone receptor 1 (PTH/PTHR1), and this interaction may mediate part of its antiosteoporotic effects. At the dose of 18 mg/kg/ day in OVX rats, genistein raised PTH levels and also dose-dependently elevated PTHR1 mRNA expression and protein content (Miao et al., 2012).

In another study by Dang et al., mRNA expression of PTH/PTHrP receptor was increased by 1 μM genistein, a concentration at which genistein has significant stimulatory effects on osteogenesis and inhibitory effects on osteoclastogenesis (Z. C. Dang et al., 2003).

Pretreatment of osteoblasts with genistein (1 μM and 100 nm) has been shown to augment the PTH-induced increase in ALP activity, to reverse the PTH-induced decrease in OPG expression, and to partially prevent PTH-induced upregulation of the RANKL:OPG ratio. Together, these results would suggest that genistein interacts with PTH and PTHR1, leading to osteogenic-promoting effects (W. F. Chen and Wong, 2006).

Currently, the main anabolic drug for osteoporosis is teriparatide (PTH 1-34), which is a recombinant form of parathyroid hormone. Genistein may be able to play a role in enhancing its antiosteoporotic actions; however, additional studies are necessary to confirm this effect.

4.4. Genistein Analogs

Several genistein analogs have been generated and described. One such compound is 6-carboxymethyl genistein, a SERM which was reported to have estrogenic activity *in vitro* and *in vivo*. The introduction of the carboxymethyl group to position 6 on genistein causes ERβ selective binding with minimal binding to ERα.

This genistein analog was demonstrated to have a stimulatory effect on ERβ and a mixed agonist/antagonist effect on ERα (Somjen et al., 2002). Wang et al. generated eight novel genistein-based SERMs, and the most promising compound was 4', 5, 7-tri [3-(2-hydroxyethylthio) propoxy] isoflavone, displaying a 24% (or 8%) increase in BMD and 31% (or 11%) increase in weight of bone ash of femora relative to those in the OVX or genistein control groups (S. F. Wang, Jiang, Ye, Li, and Tan, 2005).

Femurs were burnt to ash, and weight of bone ash was utilized for evaluating the inorganic, mineral content of the bone (S. F. Wang et al., 2005).

Zhang et al. evaluated 6-prenylgenistein, 8-prenylgenistein, and 6,8-diprenylgenistein for the effect on osteoblastic proliferation, differentiation, and mineralization. The findings demonstrated that prenylation at C-8 could increase the bone-protective effect of genistein (Zhang, Li, Yao, and Wong, 2008).

5. DAIDZEIN

5.1. Efficacy Studies of Daidzein

5.1. A. Intervention in Animals

In OVX rats treated with daidzein, daidzein treatment has the capacity to restore BMD in lumbar vertebrae (Picherit et al., 2000), femur (Picherit et al., 2000) (Mathey et al., 2007), and femoral metaphyseal (Picherit et al., 2000) (Mathey et al., 2007) and diaphyseal (Picherit et al., 2000) (Mathey et al., 2007) zones to the level of sham-operated rats. In contrast, genistein administration restored BMD only at the femoral diaphyseal zone (Picherit et al., 2000). Daidzein also preserved cancellous bone area in the distal femur metaphysis to the level of sham-operated rats (Picherit et al., 2000) (Mathey et al., 2007) and prevented the decrease in trabeculae (Mathey et al., 2007). In the study by Mathey et al., daidzein improved trabecular thickness (Mathey et al., 2007). Overall, consumption of daidzein was more efficient than consumption of genistein in preventing ovariectomy-induced bone loss in rats (Picherit et al., 2000) (Mathey et al., 2007). One possible reason for the greater efficacy of daidzein than genistein may be due to higher bioavailability of daidzein than genistein in rats and humans (Mathey et al., 2007). An additional explanation could be that daidzein was converted into its highly active metabolite equol (Mathey et al., 2007). Equol and its efficacy will be discussed in the following sections.

5.1.B. Intervention in Humans

While animal trials have shown the positive effects of soy isoflavones on preventing ovariectomy/estrogen-deficiency induced bone loss, results have been inconsistent with regard to the efficacy of daidzein in preventing bone loss in human subjects. Some studies have concluded that daidzein has no effect on bone loss when compared to placebo treatments. Other studies have concluded that daidzein has significantly increased efficacy in preventing bone loss when compared to placebo treatment groups.

Differences between individuals in ability to produce equol may be the key to explaining the variance in reported data on human health benefits of soy isoflavones, particularly daidzein (Setchell, Brown, and Lydeking-Olsen, 2002). The capacity to form equol from daidzein is suggested to be an important modulator of the response to isoflavones. Studies have shown that isoflavones have a stronger effect on putative estrogen-responsive genes in equol producers, supporting the plausibility of the importance of equol production (Niculescu, Pop, Fischer, and Zeisel, 2007).

In studies of postmenopausal women treated with isoflavones, the percent decrease in BMD (total body and regions of femur) in equol producers were less than in nonproducers (Wu et al., 2006) (Wu et al., 2007) (Ishimi, 2010). Overall, studies of soy supplementation and bone density indicate that soy products may be more effective in maintaining bone density in equol-producing individuals (Setchell et al., 2002) (Lampe, 2009).

While equol producers demonstrate better response to soy products/isoflavones, only 30-50% of humans have the intestinal microbiota necessary to metabolize daidzein into S-equol (Uehara, 2013). Gut physiology, host genetics, and diet are reported to contribute to interindividual differences in equol producing ability (Lampe, 2009) (Setchell and Cole, 2006) (Morton, Arisaka, Miyake, Morgan, and Evans, 2002).

Interestingly, interindividual variability in equol production may be unique to humans. All animals have been shown to constitutively excrete equol (Uehara, 2013) and therefore may account for studies in animal models showing less variance than human studies.

In addition to equol producer status, there are other elements which may play a role in the inconsistent reports regarding the efficacy of daidzein. These may include differences in intervention components, such as dosage, product forms (pure compound in tablet or capsule, soy extracts, isoflavone enriched soy protein, etc.), storage and processing of isoflavone (affect amount of isoflavone in final product), duration of treatment, dietary habits, the stage of osteoporosis, menopausal stage and timing of exposure to isoflavones, and bone site evaluated. Consumption of isoflavones with other foods, in various doses, and by different age groups could influence bioavailability and thus final outcome of studies (Cassidy et al., 2006). Chemical form of isoflavone may affect bioavailability as well. Bioavailability has been demonstrated to differ following consumption of fermented soybean products (containing mainly aglycones) and nonfermented soybean products (containing isoflavone glycosides) (Cassidy et al., 2006). Additionally, 6 to 18 months are needed to reach a new bone equilibrium and thus to fully demonstrate the effects of a dietary intervention on bone (Tousen et al., 2011). However, not all human interventions had a trial period of this length. Dietary studies examining efficacy of daidzein administration in humans and animals are summarized in Table 3 and Table 4, respectively. For animal dosage conversion to equivalent human doses, the Reagan-Shaw et al. study formula was used (Reagan-Shaw et al., 2008).

5.1. C. Synergism with Daidzein Treatment

The efficacy of daidzein may be increased by combining daidzein with other interventions. For example, to prevent bone loss in estrogen-deficient animals, isoflavone administration and moderate intensity exercise have been demonstrated to be more advantageous as a combined intervention than either treatment alone (Wu et al., 2006).

Fructooligosaccharides (FOS) are indigestible sugars which stimulate growth of bifidobacteria. These bacteria cleave isoflavone conjugates to yield corresponding aglycones and metabolites.

Addition of prebiotics in the form of FOS or live microbial (lactobacillus) increased the efficiency of daidzein with regard to its protective effect on femoral BMD and mechanical properties. It was suggested that this could be due to increased equol production from daidzein; however, serum equol levels were not greater than the daidzein treatment group without FOS or lactobacillus (Mathey, 2007 #145).

Table 3. Summary of studies of daidzein and equol efficacy in humans

Study Group	Duration	Isoflavone Dosage and Type	Number of subjects per treatment group	Result	References
Postmenopausal women (41-62 years)	12 months	Equol 2 mg/day, 6 mg/day, 10 mg/day	21-23	Treatment with 10 mg/ day equol prevented a decrease in total body BMD in non-equol producing postmenopausal women.	(Tousen, 2011 #141)

Table 4. Summary of studies of daidzein and equol efficacy in animals

Study Group	Duration	Isoflavone Dosage and Type	Equivalent Human Dosage	Number of subjects per treatment group	Result	References
OVX C57BL/6 mice	3 months	Daidzein (2 g Calcium + 200 mg/kg/day daidzein)	16.2162 mg/kg/day	12	High daidzein increased femur BMD but failed to increase vertebral BMD and vertebra peak load (did not preserve trabecular bone).	(Fonseca, 2004 #130)
OVX C57BL/6 mice	3 months	Equol 12 mg/day	--	15	Supplementation with equol demonstrated higher mean biomechanical values than those seen with control. Treatment with equol led to improved cortical and cancellous bone.	(Sehmisch, 2010 #61)
OVX ddY mice	2 weeks	0.06% S-equol supplemented diet (0.064 g/100 g diet)	--	9-10	Treatment with S-equol protected against ovariectomy-induced reduction in trabecular bone volume in femoral distal metaphysis and decrease in distal femur BMD.	(Nishide, 2013 #144)
OVX ddY mice	1 month	Equol 0.1 mg/day and 0.5 mg/day	--	--	0.5 mg/day equol preserved total body BMD and BMD of femur and lumbar spine.	(Fujioka, 2004 #145)
OVX Sprague Dawley rats	1.5 months	Equol 400 mg/kg/day	64.8649 mg/kg/day	10	Equol had no effect on ovariectomy-induced BMD loss at proximal tibia but decreased trabecular bone loss in the lumbar spine and increased its density.	(Rachon, 2007 #143)
OVX Female Wistar rats	3 months	Daidzein 10 micrograms/g/ day or equol 10 micrograms/g/ day	1.6216 mg/kg/day daidzein or equol	10	Femoral BMD and metaphyseal and diaphyseal femur BMD was significantly enhanced in OVX rats. Daidzein and equol demonstrated a higher protective effect than genistein on total and diaphyseal femur. Bone sparing effect of equol on whole femur was even higher than that of daidzein. Equol and daidzein treatment partially prevented the decrease of cancellous bone area and trabeculae and improved trabecular thickness. Daidzein was most efficient in improving cancellous bone area.	(Mathey, 2007 #45)
OVX Female Wistar rats	3 months	Daidzein 10 micrograms/g/ day	1.6216 mg/kg/day	13	Daidzein treatment prevented loss of BMD in lumbar vertebrae, femur, and its metaphyseal and diaphyseal zones. Daidzein treatment also prevented cancellous bone area loss in distal femur metaphysis.	(Picherit, 2000 #129)

In another study, an additive effect of dietary isoflavone conjugates and FOS was observed in the distal part of the femur and in trabecular bone (Ohta et al., 2002). These results may be due to the combination of FOS-mediated cecal β-glucosidase activity and equol production from daidzein, and thus increasing isoflavone bioavailability (Ohta et al., 2002).

The combination of daidzein and high calcium administration has been shown too favorably affect cortical and trabecular bone as indicated by femur and lumbar vertebrae BMD and biomechanical strength (Fonseca and Ward, 2004). Additionally, since isoflavones affect osteoclasts and osteoblasts directly and do not affect calcium absorption, Park et al. proposed that isoflavones may have a synergistic effect when combined with vitamin D (C. Y. Park and Weaver, 2012). It has been suggested that such a combined intervention may synergistically induce osteoblast activation and prevent osteoclast differentiation, through increase of vitamin D metabolites, vitamin D receptor (VDR), and ER actions (C. Y. Park and Weaver, 2012).

5.2. Efficacy Studies of Equol

After ingestion, isoflavone glycosides (genistin, daidzin, and glycitin) are either hydrolyzed by intestinal microflora to aglycones (genistein, daidzein, and glyceitein, respectively) or are partially absorbed in glycosidic form (Uehara, 2013). Daidzein is further metabolized predominantly to S-equol and O-desmethyl-angolensin by intestinal microflora (Lampe, 2009). Due to its more potent estrogenic effects, S-equol will be the focus of this discussion.

In comparison to daidzein, S-equol has been demonstrated to have higher estrogenicity, binding affinity, and preference to bind ERβ (Uehara, 2013). Although binding affinity of S-equol for human ERα and ERβ was found to be similar to that of genistein, equol enhanced transcription of osteogenic genes to a greater extent than genistein, particularly when combined with ERα (Morito et al., 2001). In the Hwang et al. study, among metabolites of daidzein, S-equol had the highest affinity for binding ERs, especially for binding to ERβ, and S-equol showed higher induction of transcriptional activity especially with ERβ than its precursor daidzein (Hwang et al., 2006).

5.2.A. Interventions in Animals

In OVX rats, equol has been shown to decrease bone loss in lumbar spine (Rachon, Seidlova-Wuttke, Vortherms, and Wuttke, 2007), femoral diaphysis (Mathey et al., 2007), femoral metaphysis (Mathey et al., 2007), and total femur (Mathey et al., 2007). Equol also increased the density in the lumbar spine (Rachon et al., 2007).

Equol has been shown to have no effect on ovariectomy-induced BMD loss at proximal tibia (Rachon et al., 2007). The bone sparing effect of equol on whole femur was even higher than that of daidzein, and bone strength was improved by equol consumption but not by genistein or daidzein alone (Mathey et al., 2007). These effects may be partially explained by the slow clearance rate of equol, which contributes to higher plasma concentrations in rats than those achievable with daidzein or genistein (Mathey et al., 2007).

In OVX mice, equol treatment prevented ovariectomy-induced bone loss in the femur distal metaphysis (Nishide et al., 2013), whole femur (Fujioka et al., 2004), whole lumbar spine (Fujioka et al., 2004), and whole body (Fujioka et al., 2004) (Ishimi, 2010). The study by Sehmisch et al. confirmed that equol treatment improved cortical bone density, cortical bone area, and cancellous bone density.

In addition to these parameters, equol treated mice were demonstrated to have higher mean biomechanical values than control treated mice, suggesting that equol is able to produce stronger, higher quality bone (Sehmisch et al., 2010). One mechanism by which equol may reduce estrogen deficiency related bone loss is through regulating hemopoiesis and the production of inflammatory cytokines. Nishide et al. first suggested this mechanism in the study, in which equol was observed to suppress ovariectomy-induced increase in expression of inflammatory, osteoclastogenesis, and adipogenesis related genes (Nishide et al., 2013).

With regard to 3D trabecular bone architecture, equol has demonstrated its ability to maintain architecture and prevent trabecular bone loss in OVX mice.

Specifically, equol prevented ovariectomy-induced decrease in bone volume, trabecular thickness, and trabecular number and the increase in trabecular spacing. Overall, these results indicate that equol treatment may prevent bone loss caused by estrogen deficiency (Nishide et al., 2013).

5.2.B. Interventions in Humans

Only one study administered equol alone as a dietary supplement to human subjects. Equol administration in non-equol-producing menopausal Japanese women resulted in decreased bone resorption and prevented a decrease in BMD in postmenopausal women after 12 months (Tousen, 2011 #147). While serum equol reaches peak concentration faster than daidzein, equol intake may have lower bioavailability relative to daidzein intake, as daidzein intake is able to produce equol continuously (Tousen et al., 2011).

Dietary studies examining the efficacy of equol administration in humans and animals are summarized in Table 3 and Table 4, respectively. Equol appears to have high efficacy in preventing ovariectomy-induced bone loss in animal subjects and postmenopausal bone loss in human subjects. Therefore, equol may play an important role in the inhibition of bone resorption by daidzein.

5.3. Cell Type Specific Effects

The two most common isoforms of ER are ERα and ERβ. The C-terminal ligand-binding domain and N-terminal transactivation domain of ERα and ERβ differ, which causes differences in binding affinity by phytoestrogens (Kuiper et al., 1998). Daidzein has strong transactivation activity for both ERα and ERβ, relative to estradiol, in concentrations over 100 nm, but it binds with higher affinity to ERβ than ERα (Hwang et al., 2006) (Kuiper et al., 1998). The effects of daidzein in stimulating osteogenesis and inhibiting bone resorption have been demonstrated to be mediated through ER-dependent pathways, primarily through ERβ (Somjen et al., 2011) (Salvatori, 2009 #58) (Tang, 2008 #36) (De Wilde, 2004 #85) (de Wilde, 2006 #83) (Rassi, 2002 #89). Daidzein has also been shown to have biphasic effects on osteogenesis, stimulating osteogenesis at concentrations below 20 μM and inhibiting osteogenesis at concentrations greater than 30 μM (Z. Dang and Lowik, 2004). When ERs were blocked by an inhibitor, daidzein decreased osteogenesis at concentrations higher than 20 μM, suggesting that above 20 μM, daidzein may act through ER-independent pathways to inhibit osteogenesis. Daidzein had its highest estrogenic activity at 10 μM, but its maximum stimulation of osteogenesis occurred at 20 μM, suggesting that daidzein was acting through both ER-dependent and ER-independent pathways.

Daidzein also increased PPARδ transcriptional activity, and at 20 μM daidzein, the concentration of PPARδ transcriptional activity was highest (Z. Dang and Lowik, 2004). PPARδ will be discussed further in the following sections. Overall, the biphasic effects of daidzein appear to stimulate osteogenesis through ER-dependent pathways at concentrations less than 20 μM and to inhibit osteogenesis through ER-dependent and ER-independent pathways at concentrations above 20 μM. Further study of the biphasic effects of daidzein are required to clarify the exact point of transition and also the mechanism of the ER-independent pathways. An overview of the cellular effects of daidzein is depicted in Figure 1. The effects of daidzein on preosteoclasts, preosteoblasts, and mature osteoblasts are shown in Figures 2, 5, and 7, respectively. It should be noted that the biphasic effect of daidzein at low and high doses is shown in Figure 7.

5.3.A. Osteoclasts

5.3.A.1. RANKL/RANK Pathway Effects

Similar to genistein, daidzein has been shown to inhibit the RANKL/RANK signaling pathway in osteoclast precursors, which may explain part of its role in inhibiting osteoclast maturation. Specifically, daidzein has been shown to significantly inhibit NF-κB nuclear localization (Garcia Palacios et al., 2005) and decrease NF-κB expression (Karieb and Fox, 2011). It has also been shown to inhibit NFATc1 nuclear translocation (Karieb and Fox, 2011) and decrease NFATc1 mRNA expression in a murine monocytic cell line (Tadaishi et al., 2014). Because c-Fos is the initial stimulus for cytoplasmic accumulation of NFATc1 prior to its activation and translocation into the nucleus, daidzein-induced decrease in c-Fos mRNA expression (Tadaishi et al., 2014) may be one mechanism by which daidzein inhibits NFATc1 nuclear translocation and decreases its expression. Additionally, daidzein at 10 μM concentration caused significant decreases in DC-STAMP levels in an ER independent manner; since daidzein decreases c-Fos and NFATc1 mRNA expression, the downstream target, DC-STAMP, would decrease (Karieb and Fox, 2011). Daidzein has also been shown to directly inhibit TNF-α induced bone resorption by suppressing TNF-α-induced c-Fos and NFATc1 expression through an ER-dependent pathway (Karieb and Fox, 2011). The effects of daidzein on these pathways are demonstrated in Figure 2.

Daidzein may affect the RANKL/RANK system indirectly as well. Daidzein is reported to significantly induce OPG expression at 1 nm concentration (De Wilde, Lieberherr, Colin, and Pointillart, 2004) and also at 10 nm to 1 μM (X. W. Chen et al., 2002). The mechanism of induction of OPG expression by daidzein is through an ER-dependent pathway (De Wilde et al., 2004) (X. W. Chen et al., 2002).

These effects suggest that daidzein inhibits osteoclast differentiation and maturation directly through the RANKL/RANK pathway and indirectly through stimulation of OPG production. These effects are key components of the capacity of daidzein to reduce bone resorption. An overview of the effects of daidzein on RANK, RANKL, and OPG can be seen in Figure 1.

5.3.A.2. Effects on Cell Proliferation and Apoptosis

Several studies have reported that daidzein causes a significant decrease in the number of osteoclasts or inhibits osteoclast formation, and it has been suggested that this decrease in the number of osteoclasts formed is ER-mediated (Gao and Yamaguchi, 1999b) (Rassi, Lieberherr, Chaumaz, Pointillart, and Cournot, 2002) (Tadaishi et al., 2014) (Karieb and Fox, 2011) (Rassi et al., 2002). However, several studies have also shown no evidence of the suppressive effects on osteoclast formation by daidzein (Yamagishi et al., 2001) (Hwang et al., 2006). Therefore, further studies are necessary to determine conclusively whether daidzein has an effect on the number of osteoclasts.

It has been suggested that daidzein reduction in osteoclast formation and activity is mediated via an effect on osteoclast differentiation (Karieb and Fox, 2011) (Rassi et al., 2002). Rassi et al. showed that daidzein inhibited osteoclast differentiation via increased caspase 3 expression (Rassi et al., 2002). Daidzein (10 nm concentration) promoted caspase 3 and caspase 8 cleavage, and DNA fragmentation of monocytic cells in the bone marrow. Caspase 3 cleavage was promoted through an ER-dependent pathway, suggesting that part of the effect of daidzein with regard to inhibiting osteoclast differentiation and activity is caused by increased apoptosis of osteoclast precursors mediated by ERs (Rassi et al., 2002).

5.3. A.3. Immunomodulatory Effects Relevant to Prevention of Bone Loss

Another anti-osteoclastogenic mechanism of daidzein is to reverse detrimental immune changes that occur during estrogen deficiency.

TNF-α-induced osteoclastogenesis is central to post-menopausal bone loss. Estrogen deficiency leads to an upregulation of T cells producing TNF-α, including a population of premature senescent $CD4^{+}CD28^{null}$ T cells that secrete a higher level of TNF-α and result in increased tartarate resistant acid phosphatase (TRAP) expression (Tyagi et al., 2011). In animal or cell culture, daidzein reduced ovariectomy-induced expansion of $CD4^{+}$ T cells in bone marrow, decreased TNF-α expression in $CD4^{+}$ T cells in bone marrow, increased percentage of $CD4^{+}CD28^{+}$ T cells through an ER-dependent pathway, decreased the osteoclastogenesis induced by $CD4^{+}CD28^{null}$ T cells in bone marrow, and decreased circulating TNF-α levels (Tyagi et al., 2011). Estrogen deficiency stimulates B lymphopoiesis and upregulates production of RANKL from B cells, which helps to induce osteoclastogenesis. Daidzein has been observed to down regulate B lymphopoiesis in bone marrow cells and decrease mRNA levels of RANKL in pre-B cells (Tyagi et al., 2011). Daidzein also increased OPG secretion by B cells (Tyagi et al., 2011). These effects of daidzein decreased osteoclast differentiation and thus decreased bone resorption.

5.3.B. Osteoblasts

The effects of daidzein on BMSCs undergoing osteogenic commitment are outlined below, and Figures 5 and 7 demonstrate the effects of daidzein in preosteoblasts and mature osteoblasts at varying doses.

5.3. B.1. Proliferation and Commitment

Daidzein has been demonstrated to enhance Runx2 expression at 10 μM (Karieb and Fox, 2012) and 1 nm (De Wilde et al., 2004). This effect has been observed to occur via an ER-dependent pathway (De Wilde et al., 2004). Daidzein has been shown to increase osteoblast number and the number of viable osteoblasts at concentrations ranging from 1 nm to 50 μM with the majority of studies indicating that this effect occurs at 1 μM and 10 μM (*only one study showed this ability of daidzein at 50 μM and the rest did not show this ability above 10 μM) (Karieb and Fox, 2012) (Suh et al., 2003) (Sugimoto and Yamaguchi, 2000) (Gao and Yamaguchi, 1999a) (Ge, Chen, Xie, and Zhang, 2006) (Jia, Wang, Xie, Wang, and Zhang, 2003) (Somjen et al., 2011). The effect of daidzein to increase osteoblast number has been demonstrated to be at least partially ER-mediated (Sugimoto and Yamaguchi, 2000). Additionally, daidzein inhibited apoptosis of osteoblastic cells subjected to TNF-α treatment (Suh et al., 2003). However, at high concentrations of daidzein, such as 100 μM, daidzein demonstrates an inhibitory effect on DNA synthesis and the number of viable osteoblasts (Suh et al., 2003).

5.3. B.2. Differentiation

Daidzein has been shown to increase ALP activity at a variety of concentrations ranging from 1 nm to 10 μM with the majority of the studies indicating that this effect occurs at 1 μM (Kanno et al., 2004) (Sugimoto and Yamaguchi, 2000) (Salvatori et al., 2009) (Gao and Yamaguchi, 1999a) (Ge et al., 2006) (De Wilde et al., 2004) (Jia et al., 2003). It was observed that daidzein-induced increase in ALP activity in mouse calvaria osteoblast like cells was at least partially ER-mediated (Sugimoto and Yamaguchi, 2000) (De Wilde et al., 2004). Salvatori et al. showed that daidzein was increasing ALP expression via ERβ in mature osteoblastic cells (Salvatori et al., 2009).

5.3. B.3. Maturation

Daidzein is able to enhance osteoblastic mineralization at a variety of concentrations ranging from 1 nm to 10 μM with the majority of studies demonstrating this effect at 1 μM (Karieb and Fox, 2012) (Gao and Yamaguchi, 1999a) (Kanno et al., 2004) (Ge et al., 2006) (De Wilde et al., 2004) (Chang et al., 2003). This effect has been shown to be through ER-dependent pathways (Karieb and Fox, 2012) (De Wilde et al., 2004). However, in the study conducted by Chang et al., the effects of daidzein to increase the number of mineralized nodules were not inhibited by ICI 182780, indicating that the effect of daidzein may not be completely mediated through ER-dependent pathways (Chang et al., 2003).

Daidzein has been shown to significantly enhance osteocalcin expression at concentrations between 1 nm and 10 μM with the majority of studies indicating that this effect occurs at 1 μM (Karieb and Fox, 2012) (Salvatori et al., 2009) (De Wilde et al., 2004) (Chang et al., 2003) (Jia et al., 2003). This effect has been shown to be mediated via ER-dependent pathways (De Wilde et al., 2004); however, Chang et al. demonstrated that the effects of daidzein may not be completely mediated through ER-dependent pathways (Chang, 2003 #88).

A factor of interest is IL-6. IL-6 is a local regulator of bone turnover and has been demonstrated to play a central role in bone loss observed under conditions of estrogen deficiency (Chrzan and Bradford, 2007). Daidzein had no effect on constitutive production of IL-6, IL-1β, NO, and PGE2 by osteoblastic cells (Suh et al., 2003).

However, treatment with daidzein has been shown to inhibit TNF-α, IL-1β, and hormone-mediated enhancement of IL-6 secretion by osteoblastic cells (Suh et al., 2003) (Chrzan and Bradford, 2007) (X. W. Chen et al., 2002) in an ER-dependent manner (Chrzan and Bradford, 2007) (X. W. Chen et al., 2002). Chrzan et al. suggested that daidzein may act via ERβ specifically to inhibit IL-6 secretion, since the concentration of daidzein that had the effect was 20 fold higher than the EC50 for transactivation of ERβ (13 nM) and 50 fold lower than EC50 for ERα transactivation (15.5 μM) in osteosarcoma cells.

In addition to the inhibitory effect of daidzein on IL-6, daidzein has also been shown to affect levels of another proinflammatory cytokine that plays a role in bone loss, IL-1β. In OVX mice, daidzein attenuated the increase in serum levels of IL-1β that was seen in the OVX control group (Fonseca and Ward, 2004).

5.3. C. Thyroid C Cells

Thyroid C cells produce calcitonin, which inhibits bone resorption. Calcitonin directly inhibits osteoclast activity through its receptors and causes a decrease in number and motility of osteoclasts (Filipovic et al., 2010). Daidzein has been observed to stimulate C cells and increase trabecular bone mass in the tibia of OVX rats (Filipovic et al., 2010). Daidzein also enhanced cancellous bone area, trabecular thickness, and trabecular number while reducing trabecular separation (Filipovic et al., 2010).

5.3. D. ASCs

Daidzein-treated ASCs demonstrated a 1.7 fold increase in osteogenic differentiation relative to vehicle treated ASCs (Strong et al., 2014). Daidzein-treated ASCs demonstrated increased c-FOS, COL1A1, and insulin-like growth factor 1 (IGF1) expression compared with undifferentiated ASCs (Strong et al., 2014). Daidzein effects on ASCs were inhibited by ICI 182780 (fulvestrant), indicating that the effects were in part ER-dependent (Strong et al., 2014).

5.4. Molecular Effects

Effects of daidzein on osteoblast function are primarily regulated by its rapid activation of an ERβ related membrane-bound protein at low concentrations (10-100 pM) and its activation of the classical ERβ at higher concentrations (de Wilde, Heberden, Chaumaz, Bordat, and Lieberherr, 2006).

De Wilde et al. demonstrated that low concentrations of daidzein can bind to a membrane moiety related to ERβ, and within 15 seconds, daidzein can phosphorylate ERK1/2 via two pathways: Gβ1/PLC-β2/PKC/c-Raf-1/MEK1/2 or Gβ1/PI3K/cSrc/c-Raf-1/MEK1/2 (de Wilde et al., 2006). Daidzein induced rapid (1 min) changes in actin cytoskeleton via the two pathways, and the rearrangement of the actin cytoskeleton with reorganization to the cell periphery may control osteoblast functions such as adhesion, division, secretion, and motility. Also, through MAPK ERK1/2, daidzein activated Elk1 and cAMP response element-binding protein (CREB), which are linked to early genes controlling osteoblast proliferation and differentiation (de Wilde et al., 2006). These effects of daidzein are depicted in Figure 5.

Daidzein increases intracellular calcium via two mechanisms: one involving calcium influx from the extracellular milieu and the other involving mobilization of calcium from the endoplasmic reticulum (de Wilde et al., 2006). Daidzein has been shown to enhance nuclear content of vitamin D receptors at 1 nm concentration via an ER-dependent pathway (De Wilde et al., 2004). Stimulation of pathways that lead to increased intracellular calcium levels in osteoblasts may therefore play a role in the anabolic effects of daidzein with regard to bone.

BMP2 is highly expressed in osteoblasts and is a regulator of osteogenic differentiation of MSCs, bone marrow stromal cells, and preosteoblasts. Daidzein has been shown to upregulate BMP2 expression (X. Li et al., 2009) (Jia et al., 2003). Increased BMP production in osteoblasts may help mediate the effects of daidzein on osteogenesis, resulting in increased bone formation (Jia et al., 2003). Daidzein transactivated PPARγ and also PPARδ at micromolar concentrations (Z. Dang and Lowik, 2004). Activation of PPARδ stimulated osteogenesis, whereas PPARγ inhibited osteogenesis. Daidzein-induced activation of PPARγ down regulated estrogenic transcriptional activity, whereas activation of PPARδ upregulated estrogenic transcriptional activity (Z. Dang and Lowik, 2004). Activation of ERα or ERβ by daidzein down regulated PPARγ transcriptional activity but had no impact on PPARδ transcriptional activity (Z. Dang and Lowik, 2004).

Daidzein (1 μM) modified the transcriptional activity of the EGFR gene promoter by binding both ERs, and through ER-binding, daidzein transiently inhibited EGFR mRNA expression (Salvatori et al., 2009). Daidzein also progressively reduced the EGFR protein level until it was nearly undetectable. Inhibition of EGFR led to reduced cell growth by cell cycle arrest, activation of apoptosis, and stimulation of osteogenic differentiation (Salvatori et al., 2009).

5.5. Daidzein Analogs

To identify superior phytoestrogens, groups have structurally modified daidzein. Daidzein analogs 2g and 2l were generated by modification of the 7-hydroxy moiety by substituting the hydrogen with a cyclopentyl (2g) or allyl (2l) (Strong et al., 2014). These analogs were demonstrated to increase the expression of key osteogenic genes and to induce expression of these genes at earlier time points, relative to vehicle-treated cells (Strong et al., 2014). It was demonstrated that ER antagonist fulvestrant inhibited the osteogenic differentiation of 2g in BMSCs and ASCs while fulvestrant only decreased the effects of 2l, indicating that 2l may act through ER-dependent and ER-independent pathways. At 1 μM concentration, 2l resulted in the greatest degree of osteogenic differentiation in BMSCs and ASCs, and higher doses of estradiol were unable to induce osteogenesis equal to or exceeding the levels of osteogenesis induced by 2l (Strong et al., 2014).

In a similar study conducted by Jiang et al., daidzein analogues were generated by modifying the 7-hydroxyl functional group. Structure activity relationship studies on these analogues of daidzein demonstrated a loss of ER agonist activity and a gain of ER antagonistic activity in breast cancer cells as the 7-OH hydrogen was substituted with increasingly hydrophobic groups. Short chain alkyl substitutions (compounds with a methyl, ethyl, or n-propyl substituent) weakened the estrogenic potency of daidzein but did not render compounds antiestrogenic. Longer chain alkyl substitutions (n-butyl and n-hexyl substituents) appeared to further diminish estrogenic activity.

Isopropyl, isobutyl, and cyclopentyl substituent compounds where branched alkyl or cycloalkyl substitutions were introduced, exhibited significant antiestrogenic property (Jiang et al., 2010). This suggests that compounds with cyclopentyl substituents at the 7-hydroxyl group, such as the osteogenic compound 2g from the Strong et al. study, may have antiestrogenic effects with regard to breast cancer cells.

Yadav and colleagues reported that modifying the two hydroxyl groups of daidzein, at C-7 and C-4', into alkoxy groups could lead to synthetic daidzein derivatives with altered potency (Yadav et al., 2011). One such compound, 3c (7-(2-diethylamino-ethoxy)-3-(4-methoxyphenyl)-4H-chromen-4-one), increased mineralization of bone marrow osteoprogenitor cells and increased mRNA expression of BMP2 and osteocalcin but unlike daidzein, did not inhibit osteoclastogenesis (Yadav et al., 2011). Interestingly, in addition to its increased potency with regard to its osteogenic effects, 3c demonstrated a significant anti-estrogenic effect and may be effectively used against breast cancer (Yadav et al., 2011). This supports findings of the Jiang et al. study, which found that substituting hydrophobic, bulky groups at the 7-OH group may lead to antiestrogenic effects on breast cancer cells (Jiang et al., 2010).

6. Comparison of Efficacy

6.1. Combination Intervention

Almost all studies of combination interventions with daidzein and genistein have been conducted as *in vivo* dietary supplement experiments. The studies varied in the dosage and relative amounts of the isoflavones tested, product forms (pure compound in tablet or capsule, soy extracts, isoflavone enriched soy protein, etc.), storage and processing of isoflavones, the chemical form (aglycone vs. glycoside), the duration of intervention, dietary habits of subjects, whether subjects were osteopenic/osteoporotic or general population, bone site evaluated, and various aspects of the subject tested. Consumption of isoflavones with other foods, in various doses, and by different age groups or women in different phases of menopause could influence bioavailability and final outcome of studies (Cassidy et al., 2006). Additionally, as discussed in the previous sections, the equol-producing capability of the subjects in the study adds an additional layer of variability.

There is some level of difficulty in drawing conclusions from these studies due to the inconsistency in testing conditions. Studies examining the efficacy of isoflavones under these various parameters in humans and animals are summarized in Table 5 and Table 6, respectively. For animal dosage conversion to equivalent human doses, the Reagan-Shaw et al. study formula was used (Reagan-Shaw et al., 2008).

In an *in vitro* study by Park et al., daidzein and genistein (0.1x10^-8 M each) resulted in a significant increase in secretion of OPG by osteoblasts and a decrease in secretion of RANKL (K. Park et al., 2014).

Thus, RANKL: OPG gene expression ratio significantly decreased. When the conditioned medium from these osteoblasts was used to treat osteoclasts, the RANKL-induced osteoclast formation was significantly inhibited (K. Park et al., 2014).

The daidzein/genistein combination also markedly attenuated RANKL-induced increase in expression of NFATc1, which is necessary for osteoclast differentiation (K. Park et al., 2014). To our knowledge, this is the only combination intervention with daidzein and genistein only.

It provides promising evidence that the combination of daidzein and genistein may stimulate bone formation and inhibit bone resorption.

Table 5. Summary of combined intervention efficacy studies in humans

Study Group	Duration	Isoflavone Dosage and Type	Number of subjects per treatment group	Result	References
Postmenopausal Japanese women (45-60 years)	12 months	75 mg isoflavones (47 mg aglycone form)-38.3 mg daidzin, 0.6 mg daidzein, 8.6 mg genistin, 0.2 mg genistein, 24.4 mg glycitin with glycitein, and others	25	The annualized changes in BMD in the total hip and intertrochanteric regions in isoflavone-treated equol producers were less than in the nonproducers.	(Ishimi, 2010 #135)
Postmenopausal women (45-60 years)	6 months	75 mg of isoflavone conjugates (38.3 mg daidzin, 8.6 mg genistin, 23.4 mg glycitin, and others	31-33	Percent changes in BMD in equol producers were significantly lower than changes in BMD for sub-whole body and total hip in nonproducers in isoflavone group.	(Wu, 2006 #133)
Postmenopausal Taiwanese women (45-65 years)	24 months	300 mg/day isoflavones in aglycone form (172.5 mg genistein+127.5 mg daidzein)	199-200	Isoflavones failed to prevent lumbar spine and total proximal femur BMD from declining as compared with placebo group.	(Tai, 2012 #3283)
Men and women 50-80 years old	12 months	Soy protein containing 83 mg of isoflavones in aglycone form (45.6 mg genistein, 31.7 mg daidzein)	11 women, 63 men	Soy protein containing isoflavones showed modest benefit in improving spine but not hip BMD in older women.	(Newton, 2006 #3284)
Postmenopausal Chinese women (48-62 years)	12 months	40 mg isoflavones + 500 mg calcium and 80 mg isoflavones + 500 mg calcium; isoflavone content was 46.4% daidzein, 38.8% glycetein, 14.7% genistein	68	There was a positive effect of isoflavone supplementation on rates of change of BMC at total hip and trochanter among women who were greater than four years past the onset of menopause.	(Chen, 2004 #3285)
Postmenopausal women (45-65 years)	3 months	52.64+/-8.68 mg/day daidzein, genistein, and their glycosides	--	Study participants consuming soy demonstrated a 5.2% increase in BMC, but no change in BMD was detected.	(Dalais, 1998 #3286)
Postmenopausal women (49-65 years)	12 months	26 mg biochanin A, 16 mg formononetin, 1 mg genistein, 0.5 mg daidzein	102	The isoflavone supplement demonstrated a protective effect on lumbar spine BMD and BMC with no protective effects on hip.	(Atkinson, 2004 #3287)
Postmenopausal women (mean age 70.6 +/-6.3 years)	6 months	110 mg/day isoflavones (1.3:1:0.22 ratio of genistein: daidzein: glycitein)	19	Mean spine BMD at L2 and L3 was significantly greater in isoflavone treatment group. There were nonsignificant increases from baseline for total spine BMC (3.5%), total spine BMD (1%), total hip BMC (3.6%), and total hip BMD (1.3%).	(Harkness, 2004 #3288)

Study Group	Duration	Isoflavone Dosage and Type	Number of subjects per treatment group	Result	References
Postmenopausal women (60-75 years)	12 months	25.6 g soy protein (99 mg of isoflavones including 52 mg genistein, 41 mg daidzein, and 6 mg glycetein)	75-78 in per-protocol analysis	BMD did not differ significantly between the groups.	(Kreijkamp-Kaspers, 2004 #3289)
Postmenopausal women (46-63 years)	36 months	80 mg of isoflavones or 120 mg of isoflavones/day; ratio of genistein to daidzein to glycitein (aglycone form) was 1.3:1:0.3, similar to what is found in soybeans	171 (midshaft femur) and 162 (distal tibia)	No effect was found on femur cortical thickness, periosteal circumference, or endosteal circumference. No effect was found on tibia BMD or periosteal circumference. As time since last menstrual period increased, 120 mg/day became protective of cortical BMD at the femoral neck. 80 mg/day became protective as bone turnover increased. Isoflavone treatment for 3 years was shown to be beneficial for midshaft femur vBMD as time since last menstrual period increased and for midshaft strength-strain index as bone turnover increased.	(Shedd-Wise, 2011 #3291) (Alekel, 2010 #3292)
Late postmenopausal women (>60 years)	12 months	Soy protein (18 g) + isoflavone tablets (105 mg isoflavone aglycone equivalents), soy protein + placebo tablet, control protein + isoflavones	32-34	No significant differences in BMD were observed from baseline to 12 months between groups or in BMD change between equol and non-equol producers.	(Kenny, 2009 #3228)
Premenopausal or early menopausal Chinese women (45-55 years)	6 months	4 capsules of 22.5 mg soy isoflavone/day for a total of 90 mg/day (aglycone equivalents); isoflavones were 52.2% daidzein and 47.8% genistein	40	Isoflavone supplement group demonstrated increased tibial BMD.	(Chi, 2013 #443)
Perimenopausal women	6 months	80.4 mg aglycone components/day	24	Isoflavone rich soy treatment group contributed positively to percent change in BMC and BMD in lumbar spine.	(Alekel, 2000 #3293)

However, daidzein and genistein were not tested alone in this study, so it is not possible to conclusively determine whether daidzein and genistein together had a synergistic effect. Future studies *in vivo* and *in vitro* should further explore combinations of daidzein and genistein to determine whether these phytoestrogens have a synergistic effect on protecting bone.

Table 6. Summary of combined intervention efficacy studies in animals

Study Group	Duration	Isoflavone Dosage and Type	Equivalent Human Dosage	Number of subjects per treatment group	Result	References
OVX female C57BL/6 mice	5 weeks	2500 mg/kg diet (40% isoflavones in ratio of 1.3:1:0.3 of genistein: daidzein:glycitein ratio); mice were fed 1000 mg/kg/day diet of isoflavones (Novasoy)	81.0811 mg/kg/day	12	Proximal tibial metaphysis BMD was significantly induced in response to Novasoy. Novasoy significantly preserved trabecular bone mass, bone volume, and trabecular bone separation.	(Zhang, 2009 #63)
OVX ddY mice	1.5 months	5 g/kg isoflavone with 40% isoflavones (100 mg aglycone conjugates contains 33 mg daidzein, 8.5 mg genistein, and 15 mg glycitein)	--	8	Isoflavone diet resulted in increased total BMC and distal femur BMC in OVX mice. It also resulted in increased femur metaphysis, total, cortical, and trabecular BMD.	(Ohta, 2002 #146)
OVX CD-1 mice	3 months	250 mg genistein + 250 mg daidzein/kg diet	20.2703 mg/kg/day	12	Isoflavone supplement preserved loss of total BMD in femur and femoral strength. It did not preserve BMD of lumbar vertebrae.	(Ward, 2007 #3294)
OVX Female Wistar rats	3 months	14 mg/kg/day genistein and 14 mg/kg/day daidzein (IRD)	2.2703 mg/kg/day of genistein and same amount of daidzein	7	IRD stimulated bone formation and inhibited bone resorption whereas subcutaneous or dietary administration of genistein only stimulated bone formation. IRD was more effective in bone protection than administration of pure genistein.	(Hertrampf, 2009 #49)

CONCLUSION

In vitro studies of genistein and daidzein effects to inhibit bone resorption and stimulate bone formation have been promising as have *in vivo* studies of the action of genistein to reduce phenotypic effects of osteoporosis.

Efficacy studies of daidzein in humans have shown mixed results, likely due to the role of equol production in modulating the effects of daidzein and due to the variability in study designs, making cross-comparisons and analysis difficult.

Daidzein and genistein share many cellular effects to reverse phenotypes observed in OVX animals. Shared effects to prevent resorption include inhibition of the RANKL/RANK signaling pathway, inhibition of release of proinflammatory cytokines, inhibition of activated T and B cells, and reduction of number of osteoclasts.

Shared effects to stimulate bone formation include an increase in expression of Runx2, increase in ALP activity, an upregulation in collagen expression, an increase in BMP expression, an increase in osteocalcin expression, and an increase in mineralized nodule formation. Additionally, effects unique to daidzein include its ability to induce apoptosis of osteoclasts and its ability to stimulate osteoblast proliferation and increase in osteoblast number. Effects unique to genistein include its ability to increase intracellular calcium within osteoclasts and its effect to reduce oxidative stress in cells, both of which reduce bone resorption. With regard to molecular signaling that induces these cell-specific effects, daidzein and genistein both act primarily through ER-dependent mechanisms, particularly ERβ, at lower concentrations but may act through different anti-osteogenic pathways at higher concentrations. Based on the *in vitro* and *in vivo* studies of mechanism and efficacy of daidzein and genistein, these phytoestrogens show the potential to be part of anabolic osteoporosis therapies as they effectively act to stimulate bone formation and inhibit bone resorption by a variety of mechanisms that target cellular problems observed in osteoporosis.

References

Almeida, M. (2012). Aging mechanisms in bone. *Bonekey Rep., 1*. doi: 10.1038/bonekey.2012.102.

Anderson, J. J., Ambrose, W. W. and Garner, S. C. (1998). Biphasic effects of genistein on bone tissue in the ovariectomized, lactating rat model. *Proc. Soc. Exp. Biol. Med.*, 217(3), 345-350.

Antosz, M. E., Bellows, C. G. and Aubin, J. E. (1989). Effects of transforming growth factor beta and epidermal growth factor on cell proliferation and the formation of bone nodules in isolated fetal rat calvaria cells. *J. Cell. Physiol.*, 140(2), 386-395. doi: 10.1002/jcp.1041400225.

Atteritano, M., Mazzaferro, S., Frisina, A., Cannata, M. L., Bitto, A., D'Anna, R., ... Buemi, M. (2009). Genistein effects on quantitative ultrasound parameters and bone mineral density in osteopenic postmenopausal women. *Osteoporos. Int.*, 20(11), 1947-1954. doi: 10.1007/s00198-009-0883-4.

Bandyopadhyay, A., Yadav, P. S. and Prashar, P. (2013). BMP signaling in development and diseases: a pharmacological perspective. *Biochem. Pharmacol.*, 85(7), 857-864. doi: 10.1016/j.bcp.2013.01.004.

Baron, R. and Kneissel, M. (2013). WNT signaling in bone homeostasis and disease: from human mutations to treatments. *Nat. Med.*, 19(2), 179-192.

Barrett-Connor, E., Mosca, L., Collins, P., Geiger, M. J., Grady, D., Kornitzer, M., ... Wenger, N. K. (2006). Effects of raloxifene on cardiovascular events and breast cancer in postmenopausal women. *N. Engl. J. Med.*, 355 (2), 125-137. doi: 10.1056/NEJMoa062462.

Bitto, A., Burnett, B. P., Polito, F., Marini, H., Levy, R. M., Armbruster, M. A., ... Altavilla, D. (2008). Effects of genistein aglycone in osteoporotic, ovariectomized rats: a comparison with alendronate, raloxifene and oestradiol. *Br. J. Pharmacol.*, 155(6), 896-905. doi: 10.1038/bjp.2008.305.

Bitto, A., Marini, H., Burnett, B. P., Polito, F., Levy, R. M., Irrera, N., ... Altavilla, D. (2011). Genistein aglycone effect on bone loss is not enhanced by supplemental calcium and vitamin D3: a dose ranging experimental study. *Phytomedicine*, 18(10), 879-886. doi: 10.1016/j.phymed.2011.02.003.

Black, D. M., Bauer, D. C., Schwartz, A. V., Cummings, S. R., and Rosen, C. J. (2012). Continuing bisphosphonate treatment for osteoporosis--for whom and for how long? *N. Engl. J. Med.*, 366(22), 2051-2053. doi: 10.1056/NEJMp1202623.

Bliuc, D., Nguyen, N. D., Alarkawi, D., Nguyen, T. V., Eisman, J. A., and Center, J. R. (2015). Accelerated bone loss and increased post-fracture mortality in elderly women and men. *Osteoporos. Int.* doi: 10.1007/s00198-014-3014-9.

Bliuc, D., Nguyen, N. D., Milch, V. E., Nguyen, T. V., Eisman, J. A., and Center, J. R. (2009). Mortality risk associated with low-trauma osteoporotic fracture and subsequent fracture in men and women. *Jama*, 301(5), 513-521. doi: 10.1001/jama.2009.50.

Bord, S., Horner, A., Beavan, S., and Compston, J. (2001). Estrogen receptors alpha and beta are differentially expressed in developing human bone. *J. Clin. Endocrinol. Metab.*, 86 (5), 2309-2314. doi: 10.1210/jcem.86.5.7513.

Boyce, B. F. and Xing, L. (2007). Biology of RANK, RANKL, and osteoprotegerin. *Arthritis Res. Ther.*, 9 Suppl. 1, S1. doi: 10.1186/ar2165.

Boyce, B. F., Yao, Z. and Xing, L. (2009). Osteoclasts have Multiple Roles in Bone in Addition to Bone Resorption. *Crit. Rev. Eukaryot. Gene Expr.*, 19(3), 171-180.

Boyle, W. J., Simonet, W. S. and Lacey, D. L. (2003). Osteoclast differentiation and activation. *Nature*, 423(6937), 337-342. doi: 10.1038/nature01658.

Bozec, A., Bakiri, L., Jimenez, M., Rosen, E. D., Catala-Lehnen, P., Schinke, T., ... Wagner, E. F. (2013). Osteoblast-specific expression of Fra-2/AP-1 controls adiponectin and osteocalcin expression and affects metabolism. *J. Cell. Sci.*, 126(Pt 23), 5432-5440. doi: 10.1242/jcs.134510.

Bozec, A., Bakiri, L., Jimenez, M., Schinke, T., Amling, M., and Wagner, E. F. (2010). Fra-2/AP-1 controls bone formation by regulating osteoblast differentiation and collagen production. *J. Cell. Biol.*, 190(6), 1093-1106. doi: 10.1083/jcb.201002111.

Burge, R., Dawson-Hughes, B., Solomon, D. H., Wong, J. B., King, A., and Tosteson, A. (2007). Incidence and economic burden of osteoporosis-related fractures in the United States, 2005-2025. *J. Bone Miner. Res.*, 22(3), 465-475. doi: 10.1359/jbmr.061113.

Cano, A., Dapia, S., Noguera, I., Pineda, B., Hermenegildo, C., del Val, R., ... Garcia-Perez, M. A. (2008). Comparative effects of 17beta-estradiol, raloxifene and genistein on bone 3D microarchitecture and volumetric bone mineral density in the ovariectomized mice. *Osteoporos. Int.*, 19(6), 793-800. doi: 10.1007/s00198-007-0498-6.

Capulli, M., Paone, R. and Rucci, N. (2014). Osteoblast and osteocyte: Games without frontiers. *Arch. Biochem. Biophys.*, 561(0), 3-12. doi: http://dx.doi.org/10.1016/j.abb.2014.05.003.

Cassidy, A., Albertazzi, P., Lise Nielsen, I., Hall, W., Williamson, G., Tetens, I., ... Branca, F. (2006). Critical review of health effects of soyabean phyto-oestrogens in post-menopausal women. *Proc. Nutr. Soc.*, 65(1), 76-92.

Chang, H., Jin, T. Y., Jin, W. F., Gu, S. Z., and Zhou, Y. F. (2003). Modulation of isoflavones on bone-nodule formation in rat calvaria osteoblasts in vitro. *Biomed. Environ. Sci.*, 16 (1), 83-89.

Chen, W. F. and Wong, M. S. (2006). Genistein modulates the effects of parathyroid hormone in human osteoblastic SaOS-2 cells. *Br. J. Nutr.*, 95(6), 1039-1047.

Chen, X., Garner, S. C., Quarles, L. D., and Anderson, J. J. (2003). Effects of genistein on expression of bone markers during MC3T3-E1 osteoblastic cell differentiation. *J. Nutr. Biochem.*, 14(6), 342-349.

Chen, X. W., Garner, S. C. and Anderson, J. J. (2002). Isoflavones regulate interleukin-6 and osteoprotegerin synthesis during osteoblast cell differentiation via an estrogen-receptor-dependent pathway. *Biochem. Biophys. Res. Commun.*, 295(2), 417-422.

Chien, H. H., Lin, W. L. and Cho, M. I. (2000). Down-Regulation of Osteoblastic Cell Differentiation by Epidermal Growth Factor Receptor. *Calcified Tissue International*, 67 (2), 141-150. doi: 10.1007/s00223001128.

Chrzan, B. G. and Bradford, P. G. (2007). Phytoestrogens activate estrogen receptor beta1 and estrogenic responses in human breast and bone cancer cell lines. *Mol. Nutr. Food Res.*, 51(2), 171-177. doi: 10.1002/mnfr.200600091.

Colditz, G. A., Hankinson, S. E., Hunter, D. J., Willett, W. C., Manson, J. E., Stampfer, M. J., ... Speizer, F. E. (1995). The use of estrogens and progestins and the risk of breast cancer in postmenopausal women. *N. Engl. J. Med.*, 332(24), 1589-1593. doi: 10.1056/nejm 199506153322401.

Comalada, M., Ballester, I., Bailon, E., Sierra, S., Xaus, J., Galvez, J., ... Zarzuelo, A. (2006). Inhibition of pro-inflammatory markers in primary bone marrow-derived mouse macrophages by naturally occurring flavonoids: analysis of the structure-activity relationship. *Biochem. Pharmacol.*, 72(8), 1010-1021. doi: 10.1016/j.bcp.2006.07.016.

Crisafulli, A., Altavilla, D., Squadrito, G., Romeo, A., Adamo, E. B., Marini, R., ... Squadrito, F. (2004). Effects of the phytoestrogen genistein on the circulating soluble receptor activator of nuclear factor kappaB ligand-osteoprotegerin system in early postmenopausal women. *J. Clin. Endocrinol. Metab.*, 89(1), 188-192. doi: 10.1210/jc.2003-030891.

Dai, J., Li, Y., Zhou, H., Chen, J., Chen, M., and Xiao, Z. (2013). Genistein promotion of osteogenic differentiation through BMP2/SMAD5/RUNX2 signaling. *Int. J. Biol. Sci.*, 9 (10), 1089-1098. doi: 10.7150/ijbs.7367.

Dai, R., Ma, Y., Sheng, Z., Jin, Y., Zhang, Y., Fang, L., ... Liao, E. (2008). Effects of genistein on vertebral trabecular bone microstructure, bone mineral density, microcracks, osteocyte density, and bone strength in ovariectomized rats. *J. Bone Miner. Metab.*, 26 (4), 342-349. doi: 10.1007/s00774-007-0830-4.

Dang, Z. and Lowik, C. W. (2004). The balance between concurrent activation of ERs and PPARs determines daidzein-induced osteogenesis and adipogenesis. *J. Bone Miner. Res.*, 19(5), 853-861. doi: 10.1359/jbmr.040120.

Dang, Z. C., Audinot, V., Papapoulos, S. E., Boutin, J. A., and Lowik, C. W. (2003). Peroxisome proliferator-activated receptor gamma (PPARgamma) as a molecular target for the soy phytoestrogen genistein. *J. Biol. Chem.*, 278(2), 962-967. doi: 10.1074/jbc. M209483200.

de Wilde, A., Heberden, C., Chaumaz, G., Bordat, C., and Lieberherr, M. (2006). Signaling networks from Gbeta1 subunit to transcription factors and actin remodeling via a membrane-located ERbeta-related protein in the rapid action of daidzein in osteoblasts. *J. Cell. Physiol.*, 209(3), 786-801. doi: 10.1002/jcp.20767.

De Wilde, A., Lieberherr, M., Colin, C., and Pointillart, A. (2004). A low dose of daidzein acts as an ERbeta-selective agonist in trabecular osteoblasts of young female piglets. *J. Cell. Physiol.*, 200(2), 253-262. doi: 10.1002/jcp.20008.

Dodge, J. A., Glasebrook, A. L., Magee, D. E., Phillips, D. L., Sato, M., Short, L. L., and Bryant, H. U. (1996). Environmental estrogens: effects on cholesterol lowering and bone in the ovariectomized rat. *J. Steroid Biochem. Mol. Biol.*, 59(2), 155-161.

Erlandsson, M. C., Islander, U., Moverare, S., Ohlsson, C., and Carlsten, H. (2005). Estrogenic agonism and antagonism of the soy isoflavone genistein in uterus, bone and lymphopoiesis in mice. *Apmis*, 113(5), 317-323. doi: 10.1111/j.1600-0463.2005.apm_02.x.

Fakhry, M., Hamade, E., Badran, B., Buchet, R., and Magne, D. (2013). Molecular mechanisms of mesenchymal stem cell differentiation towards osteoblasts. *World J. Stem Cells*, 5(4), 136-148. doi: 10.4252/wjsc.v5.i4.136.

Fanti, P., Monier-Faugere, M. C., Geng, Z., Schmidt, J., Morris, P. E., Cohen, D., and Malluche, H. H. (1998). The phytoestrogen genistein reduces bone loss in short-term ovariectomized rats. *Osteoporos. Int.*, 8(3), 274-281. doi: 10.1007/s001980050065.

Filipovic, B., Sosic-Jurjevic, B., Ajdzanovic, V., Brkic, D., Manojlovic-Stojanoski, M., Milosevic, V., and Sekulic, M. (2010). Daidzein administration positively affects thyroid C cells and bone structure in orchidectomized middle-aged rats. *Osteoporos. Int.*, 21(9), 1609-1616. doi: 10.1007/s00198-009-1092-x.

Fonseca, D. and Ward, W. E. (2004). Daidzein together with high calcium preserve bone mass and biomechanical strength at multiple sites in ovariectomized mice. *Bone*, 35(2), 489-497. doi: 10.1016/j.bone.2004.03.031.

Fuchs-Young, R., Glasebrook, A. L., Short, L. L., Draper, M. W., Rippy, M. K., Cole, H. W., ... Bryant, H. U. (1995). Raloxifene is a tissue-selective agonist/antagonist that functions through the estrogen receptor. *Ann. N. Y. Acad. Sci.*, 761, 355-360.

Fujioka, M., Uehara, M., Wu, J., Adlercreutz, H., Suzuki, K., Kanazawa, K., ... Ishimi, Y. (2004). Equol, a metabolite of daidzein, inhibits bone loss in ovariectomized mice. *J. Nutr.*, 134(10), 2623-2627.

Gao, Y. H. and Yamaguchi, M. (1998). Zinc enhancement of genistein's anabolic effect on bone components in elderly female rats. *Gen. Pharmacol.*, 31(2), 199-202.

Gao, Y. H. and Yamaguchi, M. (1999a). Anabolic effect of daidzein on cortical bone in tissue culture: comparison with genistein effect. *Mol. Cell. Biochem.*, 194(1-2), 93-97.

Gao, Y. H. and Yamaguchi, M. (1999b). Suppressive effect of genistein on rat bone osteoclasts: apoptosis is induced through Ca2+ signaling. *Biol. Pharm. Bull.*, 22(8), 805-809.

Garcia Palacios, V., Robinson, L. J., Borysenko, C. W., Lehmann, T., Kalla, S. E., and Blair, H. C. (2005). Negative regulation of RANKL-induced osteoclastic differentiation in RAW264.7 Cells by estrogen and phytoestrogens. *J. Biol. Chem.*, 280(14), 13720-13727. doi: 10.1074/jbc.M410995200.

Ge, Y., Chen, D., Xie, L., and Zhang, R. (2006). Enhancing effect of daidzein on the differentiation and mineralization in mouse osteoblast-like MC3T3-E1 cells. *Yakugaku Zasshi*, 126(8), 651-656.

Gercel-Taylor, C., Feitelson, A. K. and Taylor, D. D. (2004). Inhibitory effect of genistein and daidzein on ovarian cancer cell growth. *Anticancer Res.*, 24(2b), 795-800.

Gertz, E. R., Silverman, N. E., Wise, K. S., Hanson, K. B., Alekel, D. L., Stewart, J. W., ... Van Loan, M. D. (2010). Contribution of serum inflammatory markers to changes in bone mineral content and density in postmenopausal women: a 1-year investigation. *J. Clin. Densitom.*, 13(3), 277-282. doi: 10.1016/j.jocd.2010.04.003.

Greenblatt, M. B. (2010). *The p38 MAPK pathway is essential for skeletogenesis and bone homeostasis in mice.* 120(7), 2457-2473. doi: 10.1172/jci42285.

Heim, M., Frank, O., Kampmann, G., Sochocky, N., Pennimpede, T., Fuchs, P., ... Bendik, I. (2004). The phytoestrogen genistein enhances osteogenesis and represses adipogenic differentiation of human primary bone marrow stromal cells. *Endocrinology*, 145(2), 848-859. doi: 10.1210/en.2003-1014.

Hertrampf, T., Gruca, M. J., Seibel, J., Laudenbach, U., Fritzemeier, K. H., and Diel, P. (2007). The bone-protective effect of the phytoestrogen genistein is mediated via ER alpha-dependent mechanisms and strongly enhanced by physical activity. *Bone*, 40(6), 1529-1535. doi: 10.1016/j.bone.2007.02.006.

Hertrampf, T., Schleipen, B., Offermanns, C., Velders, M., Laudenbach, U., and Diel, P. (2009). Comparison of the bone protective effects of an isoflavone-rich diet with dietary and subcutaneous administrations of genistein in ovariectomized rats. *Toxicol. Lett.*, 184(3), 198-203. doi: 10.1016/j.toxlet.2008.11.006.

Hertrampf, T., Seibel, J., Laudenbach, U., Fritzemeier, K. H., and Diel, P. (2008). Analysis of the effects of oestrogen receptor alpha (ERalpha)- and ERbeta-selective ligands given in combination to ovariectomized rats. *Br. J. Pharmacol.*, 153(7), 1432-1437. doi: 10.1038/ sj.bjp.0707664.

Hu, H. M., Yang, L., Wang, Z., Liu, Y. W., Fan, J. Z., Fan, J., ... Luo, Z. J. (2013). Overexpression of integrin a2 promotes osteogenic differentiation of hBMSCs from senile osteoporosis through the ERK pathway. *Int. J. Clin. Exp. Pathol.*, 6(5), 841-852.

Huang, Y., Cao, S., Nagamani, M., Anderson, K. E., Grady, J. J., and Lu, L. J. (2005). Decreased circulating levels of tumor necrosis factor-alpha in postmenopausal women during consumption of soy-containing isoflavones. *J. Clin. Endocrinol. Metab.*, 90(7), 3956-3962. doi: 10.1210/jc.2005-0161.

Hwang, C. S., Kwak, H. S., Lim, H. J., Lee, S. H., Kang, Y. S., Choe, T. B., ... Han, K. O. (2006). Isoflavone metabolites and their in vitro dual functions: they can act as an estrogenic agonist or antagonist depending on the estrogen concentration. *J. Steroid Biochem. Mol. Biol.*, 101(4-5), 246-253. doi: 10.1016/j.jsbmb.2006.06.020.

Ishimi, Y. (2010). Dietary equol and bone metabolism in postmenopausal Japanese women and osteoporotic mice. *J. Nutr.*, 140(7), 1373s-1376s. doi: 10.3945/jn.110.124842.

Ishimi, Y., Miyaura, C., Ohmura, M., Onoe, Y., Sato, T., Uchiyama, Y., ... Ikegami, S. (1999). Selective effects of genistein, a soybean isoflavone, on B-lymphopoiesis and bone loss caused by estrogen deficiency. *Endocrinology*, 140(4), 1893-1900. doi: 10. 1210/endo.140.4.6663.

Ishimi, Y., Yoshida, M., Wakimoto, S., Wu, J., Chiba, H., Wang, X., ... Miyaura, C. (2002). Genistein, a soybean isoflavone, affects bone marrow lymphopoiesis and prevents bone loss in castrated male mice. *Bone*, 31(1), 180-185.

Jensen, E. D., Gopalakrishnan, R. and Westendorf, J. J. (2010). Regulation of Gene Expression in Osteoblasts. *Biofactors*, 36(1), 25-32. doi: 10.1002/biof.72.

Jia, T. L., Wang, H. Z., Xie, L. P., Wang, X. Y., and Zhang, R. Q. (2003). Daidzein enhances osteoblast growth that may be mediated by increased bone morphogenetic protein (BMP) production. *Biochem. Pharmacol.*, 65(5), 709-715.

Jiang, Q., Payton-Stewart, F., Elliott, S., Driver, J., Rhodes, L. V., Zhang, Q., ... Wang, G. (2010). Effects of 7-O Substitutions on Estrogenic and Antiestrogenic Activities of Daidzein Analogues in MCF-7 Breast Cancer Cells. *J. Med. Chem.*, 53(16), 6153-6163. doi: 10.1021/jm100610w.

Jin, S., Zhang, Q. Y., Kang, X. M., Wang, J. X., and Zhao, W. H. (2010). Daidzein induces MCF-7 breast cancer cell apoptosis via the mitochondrial pathway. *Ann. Oncol.*, 21(2), 263-268. doi: 10.1093/annonc/mdp499.

Ju, Y. H., Fultz, J., Allred, K. F., Doerge, D. R., and Helferich, W. G. (2006). Effects of dietary daidzein and its metabolite, equol, at physiological concentrations on the growth of estrogen-dependent human breast cancer (MCF-7) tumors implanted in ovariectomized athymic mice. *Carcinogenesis*, 27(4), 856-863. doi: 10.1093/carcin/bgi320.

Kajiya, H. (2012). Calcium signaling in osteoclast differentiation and bone resorption. *Adv. Exp. Med. Biol.*, 740, 917-932. doi: 10.1007/978-94-007-2888-2_41.

Kajiya, H., Okabe, K., Okamoto, F., Tsuzuki, T., and Soeda, H. (2000). Protein tyrosine kinase inhibitors increase cytosolic calcium and inhibit actin organization as resorbing activity in rat osteoclasts. *J. Cell. Physiol.*, 183(1), 83-90. doi: 10.1002/(sici)1097-4652 (200004)183:1<83::aid-jcp10>3.0.co;2-w.

Kalu, D. N. (1991). The ovariectomized rat model of postmenopausal bone loss. *Bone Miner.*, 15(3), 175-191.

Kanno, S., Hirano, S. and Kayama, F. (2004). Effects of phytoestrogens and environmental estrogens on osteoblastic differentiation in MC3T3-E1 cells. *Toxicology*, 196(1-2), 137-145. doi: 10.1016/j.tox.2003.12.002.

Karieb, S. and Fox, S. W. (2011). Phytoestrogens directly inhibit TNF-alpha-induced bone resorption in RAW264.7 cells by suppressing c-fos-induced NFATc1 expression. *J. Cell. Biochem.*, 112(2), 476-487. doi: 10.1002/jcb.22935.

Karieb, S. and Fox, S. W. (2012). Zinc modifies the effect of phyto-oestrogens on osteoblast and osteoclast differentiation in vitro. *Br. J. Nutr.*, 108(10), 1736-1745. doi: 10.1017/s 0007114511007355.

Karieb, S. and Fox, S. W. (2013). Suppression of T cell-induced osteoclast formation. *Biochem. Biophys. Res. Commun.*, 436(4), 619-624. doi: 10.1016/j.bbrc.2013.05.140.

Kim, J. M., Lee, S. U., Kim, Y. S., Min, Y. K., and Kim, S. H. (2008). Baicalein stimulates osteoblast differentiation via coordinating activation of MAP kinases and transcription factors. *J. Cell. Biochem.*, 104(5), 1906-1917. doi: 10.1002/jcb.21760.

Kuiper, G. G., Lemmen, J. G., Carlsson, B., Corton, J. C., Safe, S. H., van der Saag, P. T., ... Gustafsson, J. A. (1998). Interaction of estrogenic chemicals and phytoestrogens with estrogen receptor beta. *Endocrinology*, 139(10), 4252-4263. doi: 10.1210/endo.139.10. 6216.

Lamartiniere, C. A., Wang, J., Smith-Johnson, M., and Eltoum, I. E. (2002). Daidzein: bioavailability, potential for reproductive toxicity, and breast cancer chemoprevention in female rats. *Toxicol. Sci.*, 65(2), 228-238.

Lambrinoudaki, I. (2014). Progestogens in postmenopausal hormone therapy and the risk of breast cancer. *Maturitas*, 77(4), 311-317. doi: 10.1016/j.maturitas.2014.01.001.

Lampe, J. W. (2009). Is equol the key to the efficacy of soy foods? *Am. J. Clin. Nutr.*, 89(5), 1664s-1667s. doi: 10.3945/ajcn.2009.26736T.

Lee, S. H., Kim, J. K. and Jang, H. D. (2014). Genistein inhibits osteoclastic differentiation of RAW 264.7 cells via regulation of ROS production and scavenging. *Int. J. Mol. Sci.*, 15 (6), 10605-10621. doi: 10.3390/ijms150610605.

Li, B. and Yu, S. (2003). Genistein prevents bone resorption diseases by inhibiting bone resorption and stimulating bone formation. *Biol. Pharm. Bull.*, 26(6), 780-786.

Li, X., Yang, J., He, X., Yang, Z., Ding, Y., Zhao, P., ... Si, S. (2009). Identification of upregulators of BMP2 expression via high-throughput screening of a synthetic and natural compound library. *J. Biomol. Screen*, 14(10), 1251-1256. doi: 10.1177/1087057109346446.

Li, Y., Kucuk, O., Hussain, M., Abrams, J., Cher, M. L., and Sarkar, F. H. (2006). Antitumor and antimetastatic activities of docetaxel are enhanced by genistein through regulation of osteoprotegerin/receptor activator of nuclear factor-kappaB (RANK)/RANK ligand/MMP-9 signaling in prostate cancer. *Cancer Res.*, 66(9), 4816-4825. doi: 10.1158/0008-5472.can-05-3752.

Li, Y. Q., Xing, X. H., Wang, H., Weng, X. L., Yu, S. B., and Dong, G. Y. (2012). Dose-dependent effects of genistein on bone homeostasis in rats' mandibular subchondral bone. *Acta Pharmacol. Sin.*, 33(1), 66-74. doi: 10.1038/aps.2011.136.

Liao, M. H., Tai, Y. T., Cherng, Y. G., Liu, S. H., Chang, Y. A., Lin, P. I., and Chen, R. M. (2014). Genistein induces oestrogen receptor-alpha gene expression in osteoblasts through the activation of mitogen-activated protein kinases/NF-kappaB/activator protein-1 and promotes cell mineralisation. *Br. J. Nutr.*, 111(1), 55-63. doi: 10.1017/s0007114513002043.

Liao, Q. C., Xiao, Z. S., Qin, Y. F., and Zhou, H. H. (2007). Genistein stimulates osteoblastic differentiation via p38 MAPK-Cbfa1 pathway in bone marrow culture. *Acta Pharmacol. Sin.*, 28(10), 1597-1602. doi: 10.1111/j.1745-7254.2007.00632.x.

Liu, J., Xu, K., Wen, G., Guo, H., Li, S., Wu, X., ... Liao, E. (2008). Comparison of the effects of genistein and zoledronic acid on the bone loss in OPG-deficient mice. *Bone*, 42 (5), 950-959. doi: 10.1016/j.bone.2008.01.010.

Liu, X., Suzuki, N., Santosh Laxmi, Y. R., Okamoto, Y., and Shibutani, S. (2012). Anti-breast cancer potential of daidzein in rodents. *Life Sci.*, 91(11-12), 415-419.

Looker AC, B. L., Dawson-Hughes, B., Shepherd, J. A., Wright, N. C. (2012). Osteoporosis or low bone mass at the femur neck or lumbar spine in older adults: United States, 2005-2008. *NCHS data brief. no 93*. Hyattsville, MD: National Center for Health Statistics.

Manolagas, S. C. (2014). Wnt signaling and osteoporosis. *Maturitas*, 78(3), 233-237. doi: 10.1016/j.maturitas.2014.04.013.

Marie, P. J. and Kassem, M. (2011). Extrinsic mechanisms involved in age-related defective bone formation. *J. Clin. Endocrinol. Metab.*, 96(3), 600-609. doi: 10.1210/jc.2010-2113.

Marini, H., Minutoli, L., Polito, F., Bitto, A., Altavilla, D., Atteritano, M., ... Squadrito, F. (2007). Effects of the phytoestrogen genistein on bone metabolism in osteopenic postmenopausal women: a randomized trial. *Ann. Intern. Med.*, 146(12), 839-847.

Marini, H., Minutoli, L., Polito, F., Bitto, A., Altavilla, D., Atteritano, M., ... Squadrito, F. (2008). OPG and sRANKL serum concentrations in osteopenic, postmenopausal women after 2-year genistein administration. *J. Bone Miner. Res.*, 23(5), 715-720. doi: 10.1359/jbmr.080201.

Mathey, J., Mardon, J., Fokialakis, N., Puel, C., Kati-Coulibaly, S., Mitakou, S., ... Coxam, V. (2007). Modulation of soy isoflavones bioavailability and subsequent effects on bone health in ovariectomized rats: the case for equol. *Osteoporos. Int.*, 18(5), 671-679. doi: 10.1007/s00198-007-0351-y.

Matsuda, N., Yokoyama, K., Takeshita, S., and Watanabe, M. (1998). Role of epidermal growth factor and its receptor in mechanical stress-induced differentiation of human periodontal ligament cells in vitro. *Archives of Oral Biology*, 43(12), 987-997. doi: http://dx.doi.org/10.1016/S0003-9969(98)00079-X.

McLean, R. R. (2009). Proinflammatory cytokines and osteoporosis. *Curr. Osteoporos. Rep.*, 7(4), 134-139.

Miao, Q., Li, J. G., Miao, S., Hu, N., Zhang, J., Zhang, S., ... Wang, S. W. (2012). The bone-protective effect of genistein in the animal model of bilateral ovariectomy: roles of phytoestrogens and PTH/PTHR1 against post-menopausal osteoporosis. *Int. J. Mol. Sci.*, 13(1), 56-70. doi: 10.3390/ijms13010056.

Ming, L. G., Chen, K. M. and Xian, C. J. (2013). Functions and action mechanisms of flavonoids genistein and icariin in regulating bone remodeling. *J. Cell. Physiol.*, 228(3), 513-521. doi: 10.1002/jcp.24158.

Montagnani, A. (2014). Bone anabolics in osteoporosis: Actuality and perspectives. *World J. Orthop.*, 5(3), 247-254. doi: 10.5312/wjo.v5.i3.247.

Morabito, N., Crisafulli, A., Vergara, C., Gaudio, A., Lasco, A., Frisina, N., ... Squadrito, F. (2002). Effects of genistein and hormone-replacement therapy on bone loss in early postmenopausal women: a randomized double-blind placebo-controlled study. *J. Bone Miner. Res.*, 17(10), 1904-1912. doi: 10.1359/jbmr.2002.17.10.1904.

Morito, K., Hirose, T., Kinjo, J., Hirakawa, T., Okawa, M., Nohara, T., ... Masamune, Y. (2001). Interaction of phytoestrogens with estrogen receptors alpha and beta. *Biol. Pharm. Bull.*, 24(4), 351-356.

Morris, C., Thorpe, J., Ambrosio, L., and Santin, M. (2006). The soybean isoflavone genistein induces differentiation of MG63 human osteosarcoma osteoblasts. *J. Nutr.*, 136(5), 1166-1170.

Morton, M. S., Arisaka, O., Miyake, N., Morgan, L. D., and Evans, B. A. (2002). Phytoestrogen concentrations in serum from Japanese men and women over forty years of age. *J. Nutr.*, 132(10), 3168-3171.

Nakajima, D., Kim, C. S., Oh, T. W., Yang, C. Y., Naka, T., Igawa, S., and Ohta, F. (2001). Suppressive effects of genistein dosage and resistance exercise on bone loss in ovariectomized rats. *J. Physiol. Anthropol. Appl. Human Sci.*, 20(5), 285-291.

Nazrun, A. S., Tzar, M. N., Mokhtar, S. A., and Mohamed, I. N. (2014). A systematic review of the outcomes of osteoporotic fracture patients after hospital discharge: morbidity, subsequent fractures, and mortality. *Ther. Clin. Risk Manag.*, 10, 937-948. doi: 10.2147/tcrm.s72456.

Niculescu, M. D., Pop, E. A., Fischer, L. M., and Zeisel, S. H. (2007). Dietary isoflavones differentially induce gene expression changes in lymphocytes from postmenopausal women who form equol as compared with those who do not. *J. Nutr. Biochem.*, 18(6), 380-390. doi: 10.1016/j.jnutbio.2006.06.002.

Nishide, Y., Tadaishi, M., Kobori, M., Tousen, Y., Kato, M., Inada, M., ... Ishimi, Y. (2013). Possible role of S-equol on bone loss via amelioration of inflammatory indices in ovariectomized mice. *J. Clin. Biochem. Nutr.*, 53(1), 41-48. doi: 10.3164/jcbn.12-123.

Ohta, A., Uehara, M., Sakai, K., Takasaki, M., Adlercreutz, H., Morohashi, T., and Ishimi, Y. (2002). A combination of dietary fructooligosaccharides and isoflavone conjugates increases femoral bone mineral density and equol production in ovariectomized mice. *J. Nutr.*, 132(7), 2048-2054.

Okamoto, F., Okabe, K. and Kajiya, H. (2001). Genistein, a soybean isoflavone, inhibits inward rectifier K(+) channels in rat osteoclasts. *Jpn. J. Physiol.*, 51(4), 501-509.

Okumura, N., Yoshikawa, T., Iida, J., Nonomura, A., and Takakura, Y. (2006). Bone formation-promoting effect of genistein on marrow mesenchymal cell culture. *Biomed. Mater. Eng.*, 16(1), 23-32.

Onal, M., Xiong, J., Chen, X., Thostenson, J. D., Almeida, M., Manolagas, S. C., and O'Brien, C. A. (2012). Receptor activator of nuclear factor kappaB ligand (RANKL) protein expression by B lymphocytes contributes to ovariectomy-induced bone loss. *J. Biol. Chem.*, 287(35), 29851-29860. doi: 10.1074/jbc.M112.377945.

Pan, W., Quarles, L. D., Song, L. H., Yu, Y. H., Jiao, C., Tang, H. B., ... Xiao, Z. S. (2005). Genistein stimulates the osteoblastic differentiation via NO/ cGMP in bone marrow culture. *J. Cell. Biochem.*, 94(2), 307-316. doi: 10.1002/jcb.20308.

Park, C. Y. and Weaver, C. M. (2012). Vitamin D interactions with soy isoflavones on bone after menopause: a review. *Nutrients*, 4(11), 1610-1621. doi: 10.3390/nu4111610.

Park, K., Ju, W. C., Yeo, J. H., Kim, J. Y., Seo, H. S., Uchida, Y., and Cho, Y. (2014). Increased OPG/RANKL ratio in the conditioned medium of soybean-treated osteoblasts suppresses RANKL-induced osteoclast differentiation. *Int. J. Mol. Med.*, 33(1), 178-184. doi: 10.3892/ijmm.2013.1557.

Picherit, C., Coxam, V., Bennetau-Pelissero, C., Kati-Coulibaly, S., Davicco, M. J., Lebecque, P., and Barlet, J. P. (2000). Daidzein is more efficient than genistein in preventing ovariectomy-induced bone loss in rats. *J. Nutr.*, 130(7), 1675-1681.

Pie, J. E., Park, J. H., Park, Y. H., Ryu, Y. M., Kim, K. N., Suh, S. W., ... Kim, M. K. (2006). Effect of genistein on the expression of bone metabolism genes in ovariectomized mice using a cDNA microarray. *J. Nutr. Biochem.*, 17(3), 157-164. doi: 10.1016/j.jnutbio. 2005.06.002.

Power, K. A., Ward, W. E., Chen, J. M., Saarinen, N. M., and Thompson, L. U. (2006). Genistein alone and in combination with the mammalian lignans enterolactone and enterodiol induce estrogenic effects on bone and uterus in a postmenopausal breast cancer mouse model. *Bone*, 39(1), 117-124. doi: 10.1016/j.bone.2005.12.015.

Rachon, D., Seidlova-Wuttke, D., Vortherms, T., and Wuttke, W. (2007). Effects of dietary equol administration on ovariectomy induced bone loss in Sprague-Dawley rats. *Maturitas*, 58(3), 308-315. doi: 10.1016/j.maturitas.2007.09.005.

Rassi, C. M., Lieberherr, M., Chaumaz, G., Pointillart, A., and Cournot, G. (2002). Down-regulation of osteoclast differentiation by daidzein via caspase 3. *J. Bone Miner. Res.*, 17 (4), 630-638. doi: 10.1359/jbmr.2002.17.4.630.

Reagan-Shaw, S., Nihal, M. and Ahmad, N. (2008). Dose translation from animal to human studies revisited. *Faseb j.*, 22(3), 659-661. doi: 10.1096/fj.07-9574LSF.

Rickard, D. J., Monroe, D. G., Ruesink, T. J., Khosla, S., Riggs, B. L., and Spelsberg, T. C. (2003). Phytoestrogen genistein acts as an estrogen agonist on human osteoblastic cells through estrogen receptors alpha and beta. *J. Cell. Biochem.*, 89(3), 633-646. doi: 10. 1002/jcb.10539.

Rossini, M., Gatti, D. and Adami, S. (2013). Involvement of WNT/β-catenin Signaling in the Treatment of Osteoporosis. *Calcified Tissue International*, 93(2), 121-132. doi: 10.1007/s00223-013-9749-z.

Salvatori, L., Caporuscio, F., Coroniti, G., Starace, G., Frati, L., Russo, M. A., and Petrangeli, E. (2009). Down-regulation of epidermal growth factor receptor induced by estrogens and phytoestrogens promotes the differentiation of U2OS human osteosarcoma cells. *J. Cell. Physiol.*, 220(1), 35-44. doi: 10.1002/jcp.21724.

Sattui, S. E. and Saag, K. G. (2014). Fracture mortality: associations with epidemiology and osteoporosis treatment. *Nat. Rev. Endocrinol.*, 10(10), 592-602. doi: 10.1038/nrendo.2014.125.

Schairer, C., Lubin, J., Troisi, R., Sturgeon, S., Brinton, L., and Hoover, R. (2000). Menopausal estrogen and estrogen-progestin replacement therapy and breast cancer risk. *Jama*, 283(4), 485-491.

Schroder, K. (2015). NADPH oxidases in bone homeostasis and osteoporosis. *Cell. Mol. Life Sci.*, 72(1), 25-38. doi: 10.1007/s00018-014-1712-2.

Seeman, E. and Delmas, P. D. (2006). Bone quality--the material and structural basis of bone strength and fragility. *N. Engl. J. Med.*, 354(21), 2250-2261. doi: 10.1056/NEJMra053077.

Sehmisch, S., Uffenorde, J., Maehlmeyer, S., Tezval, M., Jarry, H., Stuermer, K. M., and Stuermer, E. K. (2010). Evaluation of bone quality and quantity in osteoporotic mice--the effects of genistein and equol. *Phytomedicine*, 17(6), 424-430. doi: 10.1016/j.phymed.2009.10.004.

Setchell, K. D., Brown, N. M. and Lydeking-Olsen, E. (2002). The clinical importance of the metabolite equol-a clue to the effectiveness of soy and its isoflavones. *J. Nutr.*, 132(12), 3577-3584.

Setchell, K. D. and Cole, S. J. (2006). Method of defining equol-producer status and its frequency among vegetarians. *J. Nutr.*, 136(8), 2188-2193.

Setchell, K. D. and Lydeking-Olsen, E. (2003). Dietary phytoestrogens and their effect on bone: evidence from in vitro and in vivo, human observational, and dietary intervention studies. *Am. J. Clin. Nutr.*, 78(3 Suppl.), 593s-609s.

Sinnesael, M., Claessens, F., Boonen, S., and Vanderschueren, D. (2013). Novel insights in the regulation and mechanism of androgen action on bone. *Curr. Opin. Endocrinol. Diabetes Obes.*, 20(3), 240-244. doi: 10.1097/MED.0b013e32835f7d04.

Sliwinski, L., Folwarczna, J., Janiec, W., Grynkiewicz, G., and Kuzyk, K. (2005). Differential effects of genistein, estradiol and raloxifene on rat osteoclasts in vitro. *Pharmacol. Rep.*, 57(3), 352-359.

Sliwinski, L., Folwarczna, J., Nowinska, B., Cegiela, U., Pytlik, M., Kaczmarczyk-Sedlak, I., ... Trzeciak, H. I. (2009). A comparative study of the effects of genistein, estradiol and raloxifene on the murine skeletal system. *Acta Biochim. Pol.*, 56(2), 261-270.

Small, R. E. (2005). Uses and Limitations of Bone Mineral Density Measurements in the Management of Osteoporosis. *MedGenMed*, 7(2), 3.

Somjen, D., Amir-Zaltsman, Y., Gayer, B., Kulik, T., Knoll, E., Stern, N., ... Kohen, F. (2002). 6-Carboxymethyl genistein: a novel selective oestrogen receptor modulator (SERM) with unique, differential effects on the vasculature, bone and uterus. *J. Endocrinol.*, 173(3), 415-427.

Somjen, D., Katzburg, S., Sharon, O., Grafi-Cohen, M., Knoll, E., and Stern, N. (2011). The effects of estrogen receptors alpha- and beta-specific agonists and antagonists on cell proliferation and energy metabolism in human bone cell line. *J. Cell. Biochem.*, 112(2), 625-632. doi: 10.1002/jcb.22959.

Stathopoulos, I. S., Ballas, E. G., Lampropoulou-Adamidou, K., and Trovas, G. (2014). A review on osteoporosis in men. *Hormones (Athens)*, 13(4), 441-457. doi: 10.14310/horm. 2002.1550.

Strong, A. L., Ohlstein, J. F., Jiang, Q., Zhang, Q., Zheng, S., Boue, S. M., ... Bunnell, B. A. (2014). Novel daidzein analogs enhance osteogenic activity of bone marrow-derived mesenchymal stem cells and adipose-derived stromal/stem cells through estrogen receptor dependent and independent mechanisms. *Stem Cell Res. Ther.*, 5(4), 105. doi: 10.1186/scrt493.

Sugimoto, E. and Yamaguchi, M. (2000). Anabolic effect of genistein in osteoblastic MC3T3-E1 cells. *Int. J. Mol. Med.*, 5(5), 515-520.

Suh, K. S., Koh, G., Park, C. Y., Woo, J. T., Kim, S. W., Kim, J. W., ... Kim, Y. S. (2003). Soybean isoflavones inhibit tumor necrosis factor-alpha-induced apoptosis and the production of interleukin-6 and prostaglandin E2 in osteoblastic cells. *Phytochemistry*, 63 (2), 209-215.

Tadaishi, M., Nishide, Y., Tousen, Y., Kruger, M. C., and Ishimi, Y. (2014). Cooperative effects of soy isoflavones and carotenoids on osteoclast formation. *J. Clin. Biochem. Nutr.*, 54(2), 109-115. doi: 10.3164/jcbn.13-94.

Tamama, K., Fan, V. H., Griffith, L. G., Blair, H. C., and Wells, A. (2006). Epidermal Growth Factor as a Candidate for Ex Vivo Expansion of Bone Marrow–Derived Mesenchymal Stem Cells. *Stem Cells*, 24(3), 686-695. doi: 10.1634/stemcells.2005-0176.

Tan, T. W., Huang, Y. L., Chang, J. T., Lin, J. J., Fong, Y. C., Kuo, C. C., ... Tang, C. H. (2012). CCN3 increases BMP-4 expression and bone mineralization in osteoblasts. *J. Cell. Physiol.*, 227(6), 2531-2541. doi: 10.1002/jcp.22991.

Tang, X., Zhu, X., Liu, S., Nicholson, R. C., and Ni, X. (2008). Phytoestrogens induce differential estrogen receptor beta-mediated responses in transfected MG-63 cells. *Endocrine*, 34(1-3), 29-35. doi: 10.1007/s12020-008-9099-1.

Tang, X., Zhu, X., Liu, S., Wang, S., and Ni, X. (2011). Isoflavones suppress cyclic adenosine 3',5'-monophosphate regulatory element-mediated transcription in osteoblastic cell line. *J. Nutr. Biochem.*, 22(9), 865-873. doi: 10.1016/j.jnutbio.2010.07.011.

Tella, S. H. and Gallagher, J. C. (2014). Prevention and treatment of postmenopausal osteoporosis. *J. Steroid Biochem. Mol. Biol.*, 142, 155-170. doi: 10.1016/j.jsbmb.2013. 09.008.

Tousen, Y., Ezaki, J., Fujii, Y., Ueno, T., Nishimuta, M., and Ishimi, Y. (2011). Natural S-equol decreases bone resorption in postmenopausal, non-equol-producing Japanese women: a pilot randomized, placebo-controlled trial. *Menopause*, 18(5), 563-574. doi: 10. 1097/gme.0b013e3181f85aa7.

Turner, R. T., Iwaniec, U. T., Andrade, J. E., Branscum, A. J., Neese, S. L., Olson, D. A., ... Helferich, W. G. (2013). Genistein administered as a once-daily oral supplement had no beneficial effect on the tibia in rat models for postmenopausal bone loss. *Menopause*, 20 (6), 677-686. doi: 10.1097/gme.0b013e31827d44df.

Tyagi, A. M., Srivastava, K., Sharan, K., Yadav, D., Maurya, R., and Singh, D. (2011). Daidzein prevents the increase in CD4+CD28null T cells and B lymphopoesis in ovariectomized mice: a key mechanism for anti-osteoclastogenic effect. *PLoS One*, 6(6), e21216. doi: 10.1371/journal.pone.0021216.

Uchiyama, S. and Yamaguchi, M. (2007a). Genistein and zinc synergistically enhance gene expression and mineralization in osteoblastic MC3T3-E1 cells. *Int. J. Mol. Med.*, 19(2), 213-220.

Uchiyama, S. and Yamaguchi, M. (2007b). Genistein and zinc synergistically stimulate apoptotic cell death and suppress RANKL signaling-related gene expression in osteoclastic cells. *J. Cell. Biochem.*, 101(3), 529-542. doi: 10.1002/jcb.21208.

Uehara, M. (2013). Isoflavone metabolism and bone-sparing effects of daidzein-metabolites. *J. Clin. Biochem. Nutr.*, 52(3), 193-201. doi: 10.3164/jcbn.13-2.

Viereck, V., Grundker, C., Blaschke, S., Siggelkow, H., Emons, G., and Hofbauer, L. C. (2002). Phytoestrogen genistein stimulates the production of osteoprotegerin by human trabecular osteoblasts. *J. Cell. Biochem.*, 84(4), 725-735.

Wagner, E. F. (2010). Bone development and inflammatory disease is regulated by AP-1 (Fos/Jun). *Annals of the Rheumatic Diseases*, 69(Suppl. 1), i86-i88. doi: 10.1136/ard.2009.119396.

Wang, S. F., Jiang, Q., Ye, Y. H., Li, Y., and Tan, R. X. (2005). Genistein derivatives as selective estrogen receptor modulators: sonochemical synthesis and in vivo anti-osteoporotic action. *Bioorg. Med. Chem.*, 13 (16), 4880-4890. doi: 10.1016/j.bmc.2005.04.082.

Wang, Z. L., Sun, J. Y., Wang, D. N., Xie, Y. H., Wang, S. W., and Zhao, W. M. (2006). Pharmacological studies of the large-scaled purified genistein from Huaijiao (Sophora japonica-Leguminosae) on anti-osteoporosis. *Phytomedicine*, 13(9-10), 718-723. doi: 10.1016/j.phymed.2005.09.005.

Weitzmann, M. N. (2014). T-cells and B-cells in osteoporosis. *Curr. Opin. Endocrinol. Diabetes Obes.*, 21(6), 461-467. doi: 10.1097/med.0000000000000103.

Weitzmann, M. N. and Pacifici, R. (2006a). Estrogen deficiency and bone loss: an inflammatory tale. *J. Clin. Invest.*, 116(5), 1186-1194. doi: 10.1172/jci28550.

Weitzmann, M. N. and Pacifici, R. (2006b). Estrogen regulation of immune cell bone interactions. *Ann. N. Y. Acad. Sci.*, 1068, 256-274. doi: 10.1196/annals.1346.030.

Wu, J., Oka, J., Ezaki, J., Ohtomo, T., Ueno, T., Uchiyama, S., ... Ishimi, Y. (2007). Possible role of equol status in the effects of isoflavone on bone and fat mass in postmenopausal Japanese women: a double-blind, randomized, controlled trial. *Menopause*, 14(5), 866-874. doi: 10.1097/gme.0b013e3180305299.

Wu, J., Oka, J., Higuchi, M., Tabata, I., Toda, T., Fujioka, M., ... Ishimi, Y. (2006). Cooperative effects of isoflavones and exercise on bone and lipid metabolism in postmenopausal Japanese women: a randomized placebo-controlled trial. *Metabolism*, 55 (4), 423-433. doi: 10.1016/j.metabol.2005.10.002.

Wu, J., Wang, X. X., Chiba, H., Higuchi, M., Takasaki, M., Ohta, A., and Ishimi, Y. (2003). Combined intervention of exercise and genistein prevented androgen deficiency-induced bone loss in mice. *J. Appl. Physiol.* (1985), 94(1), 335-342. doi: 10.1152/japplphysiol.00498.2002.

Wu, J., Wang, X. X., Takasaki, M., Ohta, A., Higuchi, M., and Ishimi, Y. (2001). Cooperative effects of exercise training and genistein administration on bone mass in ovariectomized mice. *J. Bone Miner. Res.*, 16(10), 1829-1836. doi: 10.1359/jbmr.2001.16.10.1829.

Yadav, D. K., Gautam, A. K., Kureel, J., Srivastava, K., Sahai, M., Singh, D., ... Maurya, R. (2011). Synthetic analogs of daidzein, having more potent osteoblast stimulating effect. *Bioorg. Med. Chem. Lett.*, 21(2), 677-681. doi: 10.1016/j.bmcl.2010.12.008.

Yamagishi, T., Otsuka, E. and Hagiwara, H. (2001). Reciprocal control of expression of mRNAs for osteoclast differentiation factor and OPG in osteogenic stromal cells by genistein: evidence for the involvement of topoisomerase II in osteoclastogenesis. *Endocrinology*, 142(8), 3632-3637. doi: 10.1210/endo.142.8.8310.

Yamaguchi, M. (2012). Nutritional factors and bone homeostasis: synergistic effect with zinc and genistein in osteogenesis. *Mol. Cell. Biochem.*, 366(1-2), 201-221. doi: 10.1007/s 11010-012-1298-7.

Yamaguchi, M. and Gao, Y. H. (1998). Anabolic effect of genistein and genistin on bone metabolism in the femoral-metaphyseal tissues of elderly rats: the genistein effect is enhanced by zinc. *Mol. Cell. Biochem.*, 178(1-2), 377-382.

Yamaguchi, M., Gao, Y. H. and Ma, Z. J. (2000). Synergistic effect of genistein and zinc on bone components in the femoral-metaphyseal tissues of female rats. *J. Bone Miner. Metab.*, 18(2), 77-83.

Yamaguchi, M. and Weitzmann, M. N. (2009). The estrogen 17beta-estradiol and phytoestrogen genistein mediate differential effects on osteoblastic NF-kappaB activity. *Int. J. Mol. Med.*, 23(2), 297-301.

Yang, L., Yu, Z., Qu, H., and Li, M. (2014). Comparative effects of hispidulin, genistein, and icariin with estrogen on bone tissue in ovariectomized rats. *Cell Biochem. Biophys.*, 70 (1), 485-490. doi: 10.1007/s12013-014-9945-0.

Zhang, Y., Li, Q., Wan, H. Y., Helferich, W. G., and Wong, M. S. (2009). Genistein and a soy extract differentially affect three-dimensional bone parameters and bone-specific gene expression in ovariectomized mice. *J. Nutr.*, 139(12), 2230-2236. doi: 10.3945/jn. 109.108399.

Zhang, Y., Li, X. L., Yao, X. S., and Wong, M. S. (2008). Osteogenic activities of genistein derivatives were influenced by the presence of prenyl group at ring A. *Arch. Pharm. Res.*, 31(12), 1534-1539. doi: 10.1007/s12272-001-2147-5.

INDEX

B

C

D

E

F

M

Q

R

S

T

U

V

W

Y

Z

α

β

γ